Water Quality of Streams Draining Abandoned and Reclaimed Mined Lands in the Kantishna Hills Area, Denali National Park and Preserve, Alaska, 2008–11

By Timothy P. Brabets and Robert T. Ourso

Prepared in cooperation with the National Park Service

Scientific Investigations Report 2013-5048

U.S. Department of the Interior
U.S. Geological Survey

U.S. Department of the Interior
KEN SALAZAR, Secretary

U.S. Geological Survey
Suzette M. Kimball, Acting Director

U.S. Geological Survey, Reston, Virginia: 2013

For more information on the USGS—the Federal source for science about the Earth, its natural and living resources, natural hazards, and the environment, visit http://www.usgs.gov or call 1–888–ASK–USGS.

For an overview of USGS information products, including maps, imagery, and publications, visit http://www.usgs.gov/pubprod

To order this and other USGS information products, visit http://store.usgs.gov

Suggested citation:
Brabets, T.P., and Ourso, R.T., 2013, Water quality of streams draining abandoned and reclaimed mined lands in the Kantishna Hills area, Denali National Park and Preserve, Alaska, 2008–11: U.S. Geological Survey Scientific Investigations Report 2013–5048, 72 p.

Contents

Contents

Figures

Figures

Tables

Conversion Factors, Datums, and Abbreviations and Acronyms

Conversion Factors

Inch/Pound to SI

Multiply	By	To obtain
Length		
inch (in.)	2.54	centimeter (cm)
inch (in.)	25.4	millimeter (mm)
foot (ft)	0.3048	meter (m)
mile (mi)	1.609	kilometer (km)
Area		
acre	4,047	square meter (m^2)
acre	0.004047	square kilometer (km^2)
square foot (ft^2)	0.09290	square meter (m^2)
square mile (mi^2)	259.0	hectare (ha)
square mile (mi^2)	2.590	square kilometer (km^2)
Volume		
gallon (gal)	3.785	liter (L)
gallon (gal)	0.003785	cubic meter (m^3)
gallon (gal)	3.785	cubic decimeter (dm^3)
cubic foot (ft^3)	0.02832	cubic meter (m^3)
cubic yard (yd^3)	0.7646	cubic meter (m^3)
Flow Rate		
foot per second (ft/s)	0.3048	meter per second (m/s)
cubic foot per second (ft^3/s)	0.02832	cubic meter per second (m^3/s)
cubic foot per second per square mile [(ft^3/s)/mi^2]	0.01093	cubic meter per second per square kilometer [(m^3/s)/km^2]
Mass		
ounce, avoirdupois (oz)	28.35	gram (g)
pound, avoirdupois (lb)	0.4536	kilogram (kg)
ton, short (2,000 lb)	0.9072	megagram (Mg)
Hydraulic gradient		
foot per mile (ft/mi)	0.1894	meter per kilometer (m/km)

SI to Inch/Pound

Multiply	By	To obtain
Length		
centimeter (cm)	0.3937	inch (in.)
millimeter (mm)	0.03937	inch (in.)
micrometer (μm)	0.000039370	inch (in.)
Mass		
gram (g)	0.03527	ounce, avoirdupois (oz)

Conversion Factors, Datums, and Abbreviations and Acronyms

Temperature in degrees Celsius (°C) may be converted to degrees Fahrenheit (°F) as follows:

$$°F = (1.8 \times °C) + 32.$$

Temperature in degrees Fahrenheit (°F) may be converted to degrees Celsius (°C) as follows:

$$°C = (°F - 32)/1.8.$$

Specific conductance is given in microsiemens per centimeter at 25 degrees Celsius (µS/cm at 25 °C).

Concentrations of chemical constituents in water are given either in milligrams per liter (mg/L) or micrograms per liter (µg/L).

Datums

Vertical coordinate information is referenced to the North American Vertical Datum of 1988 (NAVD 88).

Horizontal coordinate information is referenced to the North American Datum of 1983 (NAD 83).

Elevation, as used in this report, refers to distance above the vertical datum.

Abbreviations and Acronyms

ADAS	Algal Data Analysis System
ANILCA	Alaska National Interest Lands Conservation Act
CCME	Canadian Council of Ministers of the Environment
DENA	Denali National Park and Preserve
DOC	Dissolved Organic Carbon
EPT	Ephemeroptera (mayfly), Plecotera (stonefly), and Trichoptera (caddisfly)
FNU	Formazin Nephelometric Units
IDAS	Invertebrate Data Analysis System
NAWQA	National Water-Quality Assessment Program
NMDS	non-metric multidimensional scaling
NWQL	National Water-Quality Laboratory
PEC	Probable Effect Concentration
PEL	Probable Effect Level
QMH	Qualitative Multi-Habitat
RTH	Richest Targeted Habitat
USGS	U.S. Geological Survey
WRST	Wrangell-St. Elias National Park and Preserve

Water Quality of Streams Draining Abandoned and Reclaimed Mined Land in the Kantishna Hills Area, Denali National Park and Preserve, Alaska, 2008–11

By Timothy P. Brabets and Robert T. Ourso

Abstract

The Kantishna Hills are an area of low elevation mountains in the northwest part of Denali National Park and Preserve, Alaska. Streams draining the Kantishna Hills are clearwater streams that support several species of fish and are derived from rain, snowmelt, and subsurface aquifers. However, the water quality of many of these streams has been degraded by mining. Past mining practices generated acid mine drainage and excessive sediment loads that affected water quality and aquatic habitat. Because recovery through natural processes is limited owing to a short growing season, several reclamation projects have been implemented on several streams in the Kantishna Hills region. To assess the current water quality of streams in the Kantishna Hills area and to determine if reclamation efforts have improved water quality, a cooperative study between the U.S. Geological Survey and the National Park Service was undertaken during 2008–11.

High levels of turbidity, an indicator of high concentrations of suspended sediment, were documented in water-quality data collected in the mid-1980s when mining was active. Mining ceased in 1985 and water-quality data collected during this study indicate that levels of turbidity have declined significantly. Turbidity levels generally were less than 2 Formazin Nephelometric Units and suspended sediment concentrations generally were less than 1 milligram per liter during the current study. Daily turbidity data at Rock Creek, an unmined stream, and at Caribou Creek, a mined stream, documented nearly identical patterns of turbidity in 2009, indicating that reclamation as well as natural revegetation in mined streams has improved water quality.

Specific conductance and concentrations of dissolved solids and major ions were highest from streams that had been mined. Most of these streams flow into Moose Creek, which functions as an integrator stream, and dilutes the specific conductance and ion concentrations. Calcium and magnesium are the dominant cations, and bicarbonate and sulfate are the dominant anions. Water samples indicate that the water from Rock Creek, Moose Creek, Slate Creek, and Eldorado Creek is a calcium bicarbonate-type water. The remaining sites are a calcium sulfate type water.

U.S. Environmental Protection Agency guidelines for arsenic and antimony in drinking water were exceeded in water at Slate Creek and Eureka Creek. Concentrations of arsenic, cadmium, chromium, copper, lead, nickel, and zinc in streambed sediments at many sites exceed sediment quality guideline thresholds that could be toxic to aquatic life. However, assessment of these concentrations, along with the level of organic carbon detected in the sediment, indicate that only concentrations of arsenic and chromium may be toxic to aquatic life at many sites.

In 2008 and 2009, 104 macroinvertebrate taxa and 164 algae taxa were identified from samples collected from seven sites. Of the macroinvertebrates, 86 percent were insects and most of the algae consisted of diatoms. Based on the National Community Index, Rock Creek, a reference site, and Caribou Creek, and a mined stream that had undergone some reclamation, exhibited the best overall stream conditions; whereas Slate Creek and Friday Creek, two small streams that were mined extensively, exhibited the worst stream conditions. A non-metric multi-dimensional scaling analysis of the macroinvertebrate and algae data showed a distinct grouping between the 2008 and 2009 samples, likely because of differences between a wet, cool summer in 2008 and a dry, warm summer in 2009.

Introduction

The Kantishna Hills are in the northwest part of Denali National Park and Preserve (DENA), Alaska (fig. 1). Gold was first discovered in the northern Kantishna Hills in the stream gravels of Chitsia Creek in 1903, and the first commercial discoveries of gold in the Kantishna Hills occurred in 1904 in Friday and Eureka Creeks (fig. 2). A stampede of several thousand placer miners followed, but most miners had left by 1910, as it became evident that only a few streams carried profitable quantities of gold. Prospecting for lode deposits continued, and reached a peak in 1916, when numerous lode prospects were located along a 40-mi corridor from Slate Creek to Stampede Creek (National Park Service, 1990).

A.

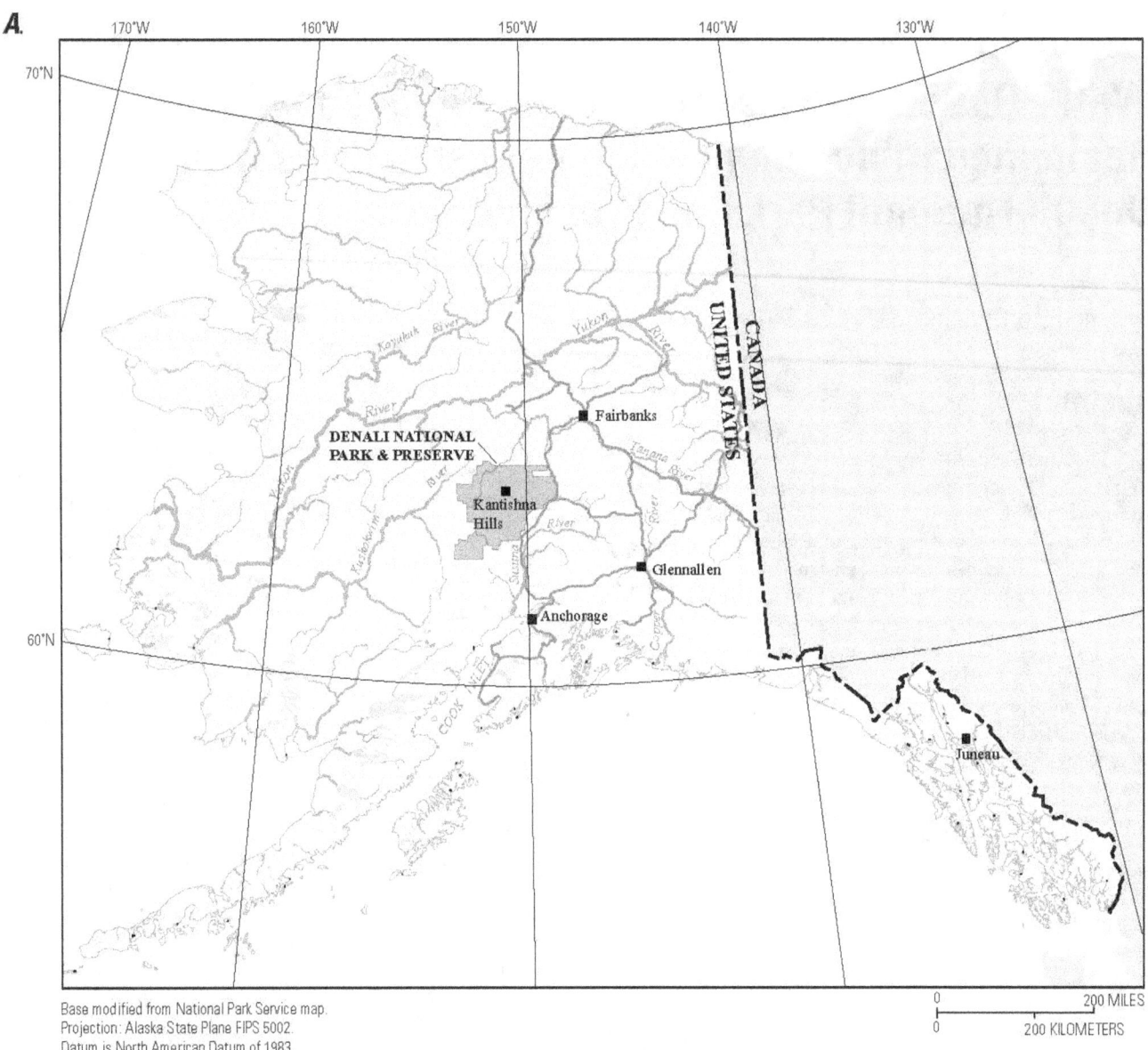

Base modified from National Park Service map.
Projection: Alaska State Plane FIPS 5002.
Datum is North American Datum of 1983.

Figure 1.   Location of Denali National Park and Preserve and Kantishna Hills, Alaska.

B.

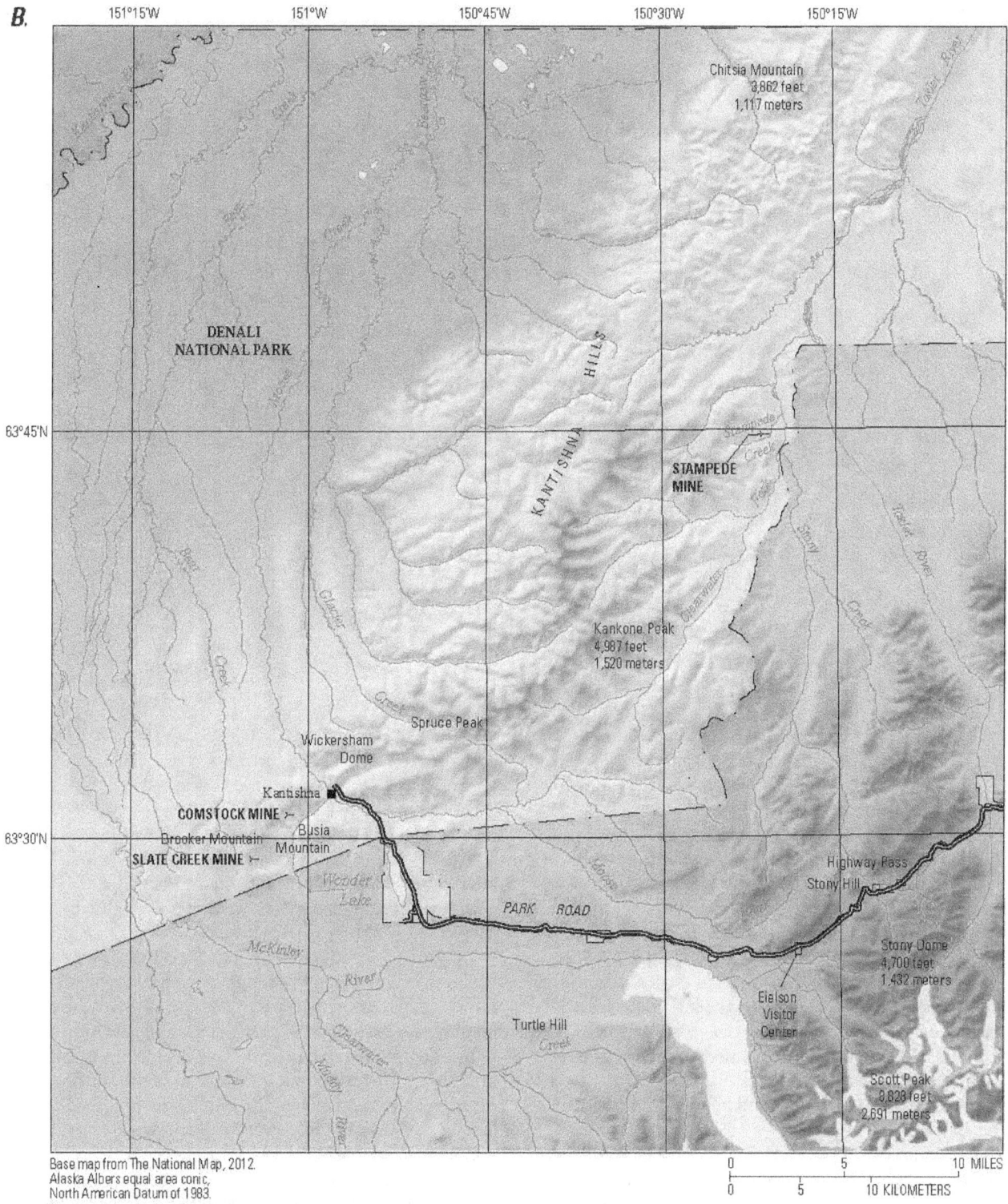

Base map from The National Map, 2012.
Alaska Albers equal area conic,
North American Datum of 1983.

Figure 1.—Continued

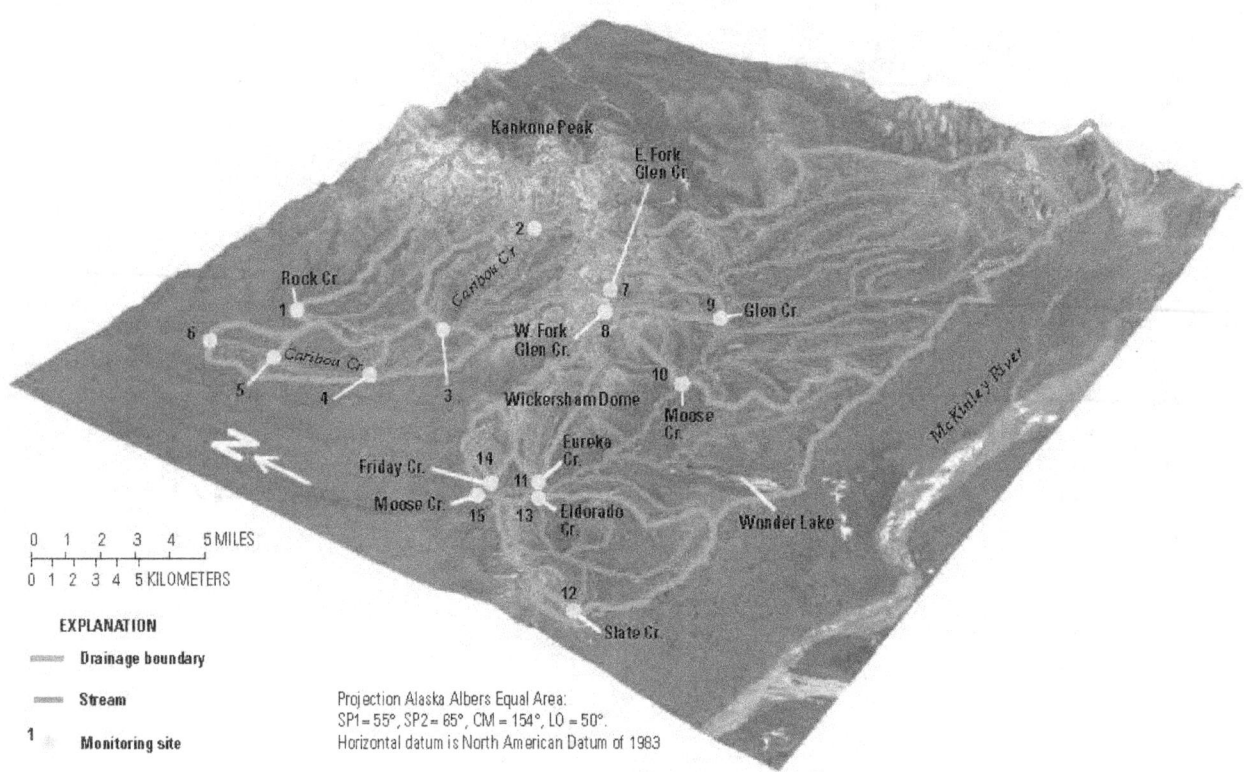

Figure 2. Kantishna Hills, Alaska.

The 1930s were considered the golden era for mining in the Kantishna mining district. In 1938, the Alaska Road Commission completed the road from the Alaska Railroad Station (located at the entrance to Denali National Park) to Kantishna Mining District. This important milestone enabled draglines, bulldozers, and other heavy equipment to be transported and used for mining. The Stampede Mine began producing antimony ore in 1936, and by 1941, was the largest antimony producing mine in Alaska. The Banjo Mine on Wickersham Dome began producing lode gold in 1939, and became the largest lode gold producer in the Kantishna mining district's history. Large-scale placer mining was successfully reintroduced in the district in the late 1930s, when firms such as Caribou Mines began operating mechanical draglines on Caribou Creek. Annual gold production in the district reached an all time high of 7,000 oz of gold in 1940 (National Park Service, 1990).

Following World War II and the eventual deregulation of the price of gold in 1972, placer mining activity resurged in the area. By 1983, total mineral production in Kantishna was estimated at 85,000 oz of gold, 265,000 oz of silver,

504,000 lbs of lead, 4,400,000 lbs of antimony, and several million pounds of combined lead and zinc (Thornsberry and others, 1984). In 1980, after the Alaska National Interest Lands Conservation Act (ANILCA) was passed, DENA tripled in size and the Kantishna Mining District was incorporated into DENA. Because mining claims were now within DENA, they were subject to the Mining in the Parks Act of 1976. As a result of a lawsuit and subsequent injunction, all mining was stopped in 1985.

Mining operations in the Kantishna Hills resulted in changes to the natural hydrologic system by stripping the overburden material to expose the mineral-bearing materials, modifying or rerouting of stream channels, sluicing or separating of gold or other recoverable minerals from the mineral-bearing materials, construction of settling ponds or other control structures, and disposal of tailings. The addition of sediment into stream systems resulted in increased turbidity, alterations to channel slope, stream velocity, and deposition of fine material and gravel on riffle areas. Changes in stream bottom material such as the particle-size composition changed the rate of intergravel water flow.

Aquatic life (benthic invertebrates and algae) was affected by reduction in photosynthetic activity and smothering of plant life on the stream bottom. Other water quality issues resulting from mining operations (Madison, 1981) could have been an increase in organic loading from the introduction of overburden materials, an increase in trace element concentrations as the result of exposure and oxidation of metal-bearing materials, the leaching of tailings deposits, or chemical treatment of the ores. Acid mine drainage—when sulfide minerals (most commonly iron sulfide or pyrite) are exposed to weathering or oxidation—also may have been present.

Currently (2012), the effects of mining are visible on more than 1,500 acres of land covering 12 watersheds in the Kantishna Hills area (National Park Service, 2001). Since 1990, DENA has acquired numerous mining claim lands, where mine waste, such as drums of hazardous material drums and lead-acid batteries have been abandoned. These hazardous materials pose a risk of entering the surface and subsurface soils if left to corrode on-site. Localized mining effects in the Kantishna Hills area have caused increased turbidity and suspended sediment loads in streams, as well as heavy metal contamination (National Park Service, 2003). Two streams, Caribou Creek and Slate Creek were included in the Alaska U.S. Environmental Protection Agency (USEPA) Category 5, Section 303(d) Listed Waterbodies for suspended sediment (Alaska Department of Environmental Conservation, 2006).

As DENA continues to acquire former mining claims where substantial environmental disturbance occurred, well-planned and comprehensive reclamation efforts are needed. Past studies indicated that riparian ecosystem recovery through natural processes can be significantly hindered because of the extensive disturbances to riparian areas caused by mining, such as vegetation and topsoil removal or floodplain and streambed excavations (Karle and Densmore, 1994). In 1988, DENA began comprehensive research on methods to promote and facilitate riparian ecosystem recovery from the effects of mining (National Park Service, 2001). The lower part of Glen Creek has been the focus of DENA efforts in floodplain restoration techniques for 10 years (Karle and Densmore, 1994). Subsequent reclamation efforts have included Slate Creek (1997), Lower Eureka Creek (1998–2000), Friday Creek (1993, 1999–2000), and Caribou Creek (2001–03). Various reclamation methods have included placing limestone rock to buffer acid drainage, capping contaminated soils, stabilizing streambanks, and revegetating disturbed areas. In 2001, the National Park Service published an environmental assessment on a proposed 10-year plan to reclaim and restore approximately 517 acres of floodplain and wetlands disturbed by mining in 12 watersheds in the Kantishna Hills (National Park Service, 2001). Reclamation has been identified by DENA as a high priority item (Mangi Environmental Group, 2005).

Purpose and Scope

This report describes the current water-quality and biology characteristics of streams in the Kantishna Hills. The scope of this report focuses on streams in the Kantishna Hills that have been affected by mining, with the exception of one stream (Rock Creek) that serves as a reference site. A wide range of samples were collected from 2008 to 2011 and analyzed for various physical, chemical, and biological constituents. In addition to describing the current water quality conditions, these data also were compared to historical data collected from these same streams in the mid-1980s and mid-1990s to determine if water-quality conditions have improved—either naturally or by reclamation efforts.

Physical Setting

The Kantishna Hills are low elevation mountains located within the continental climate zone of interior Alaska, which is characterized by seasonal temperature extremes driven chiefly by continental cooling influences and large variations in solar radiation (Shulski and Wendler, 2007). The area has short, warm summers and long, extremely cold winters. The continental pattern is modified by the higher elevations, resulting in cooler summers and higher precipitation. Streams draining from the Kantishna Hills are derived from rain, snowmelt, and subsurface aquifers. Precipitation averages 19 in. annually, with 72 percent occurring from June to September. Snow is generally present from October to May and snow accumulation ranges from 20 to 60 in. Though the annual precipitation is considered moderate, it rains during about 50 percent of the summer days. Periods of warm, dry weather extending for 1 week or more are likely to occur in June and early July. Soil temperatures are generally cold and discontinuous permafrost may be present throughout the Kantishna Hills area.

Based on records from the Wonder Lake weather station (Western Regional Climate Center, 2012), during 2008–11 total rainfall from May 1 to September 30 ranged from 6.65 in. (2009) to 14.7 in. (2008) (fig. 3). Average daily air temperature ranged from 46 degrees Fahrenheit (°F) (2008) to 49°F (2010) (fig. 4). The most notable climate features during 2008–11 were above average total rainfall in 2008, below average total rainfall in 2009, and above average air temperature in July 2009 (60°F).

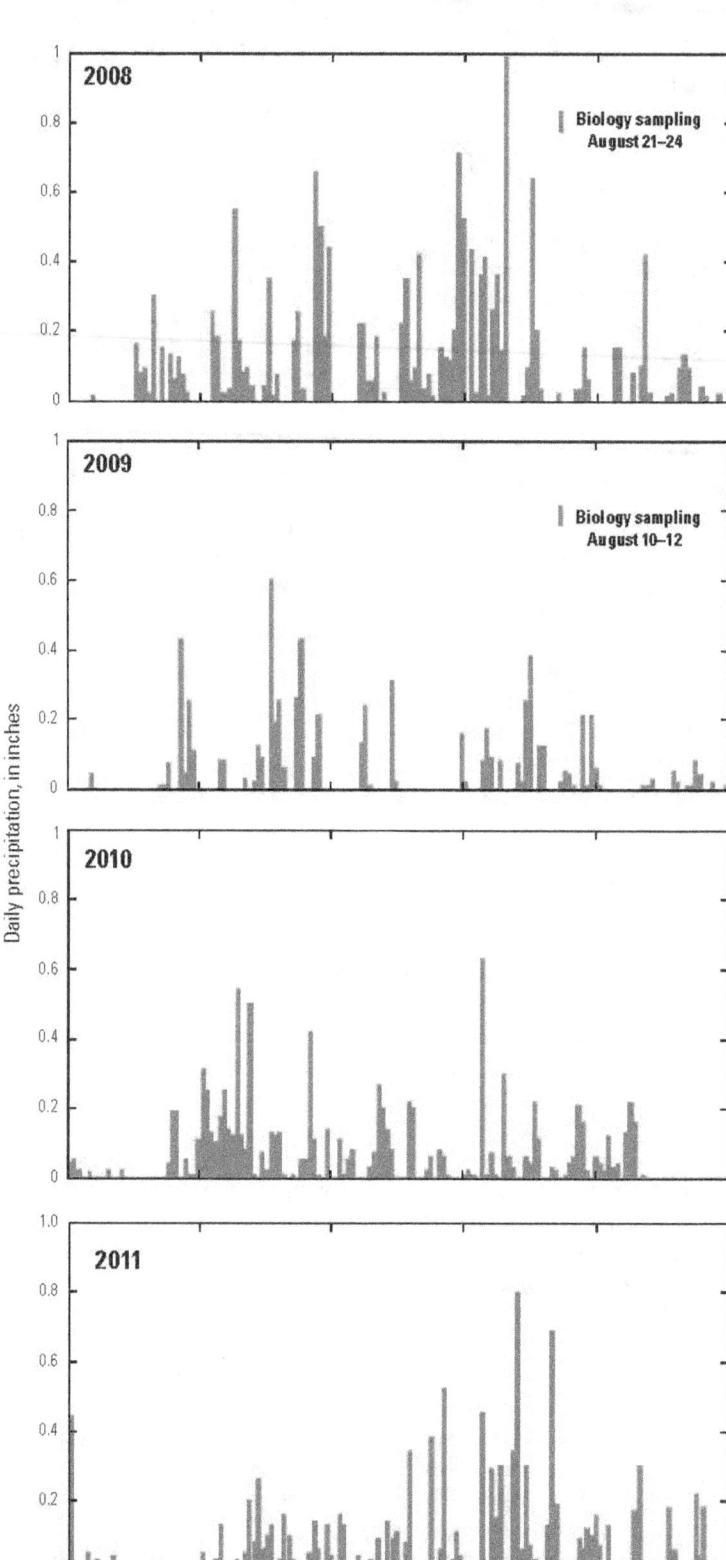

Figure 3.   Daily precipitation recorded at Wonder Lake, Alaska, May 1 to September 30, 2008–11.

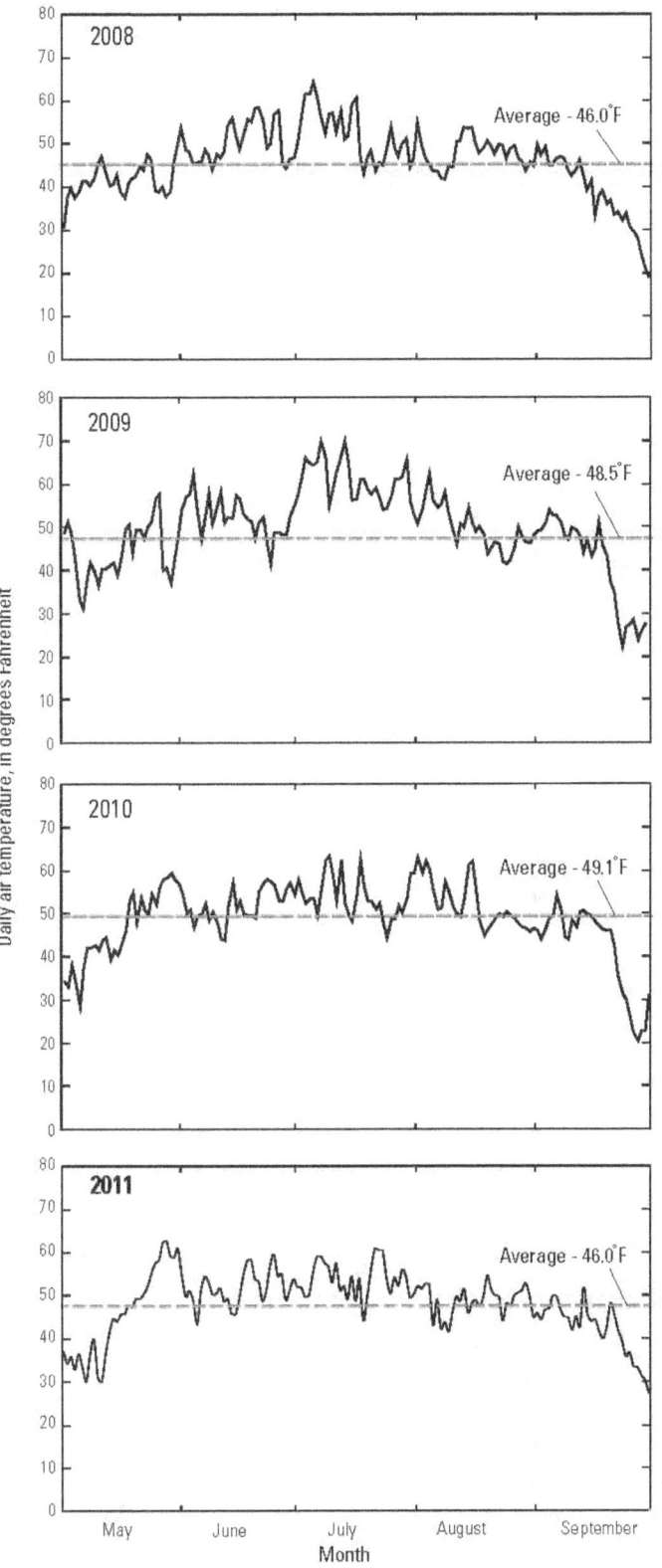

Figure 4.   Average daily air temperature at Wonder Lake, Alaska, May 1 to September 30, 2008–11.

Bedrock in the Kantishna Hills area is regionally metamorphosed, ranges in age from Precambrian to late Paleozoic, and consists of four principle rock units. The oldest and most spatially extensive rocks are Precambrian metamorphic rocks, composed of quartzite, quartz-mica schist, feldsparbiotite schist, gneiss, marble, and greenstone, called the Birch Creek Schist by Bundtzen (1981). The remaining three principle rock units, the Spruce Creek sequence, the Keevy Peak formation, and the Totatlanika Schist, are regionally metamorphosed to a lower grade than, and tectonically juxtaposed against, the Precambrian metamorphic rocks (Bundtzen, 1981).

The Kantishna Hills are at the tree line, and trees are confined to favorable sites on alluvial terraces and south-facing slopes. Tall shrubs dominate riparian vegetation on the floodplain and younger terraces, and low shrubs and herbs form the tundra vegetation on colder, more exposed sites. Mining severely disturbed the vegetation in the study area, but the pre-disturbance vegetation can be inferred from remnants and adjacent undisturbed watersheds. In the undisturbed watersheds, the floodplain is dominated by 10–13 ft tall feltleaf willow (*Salix alaxensis*) (Viereck and Little, 1972), mixed with varying amounts of 3–6 ft tall American green alder (Hulten, 1968). Natural flood or ice events, which remove vegetation and initiate primary riparian succession, seem to be infrequent. Higher elevation areas have balsam poplar and younger white spruce, and old terraces have open stands of white spruce with an understory of low shrubs, including dwarf birch and diamondleaf willow.

Watershed Specifics and Mining History

Specific information on the watersheds studied and mining history is given below (fig. 2, fig. 5, table 1).

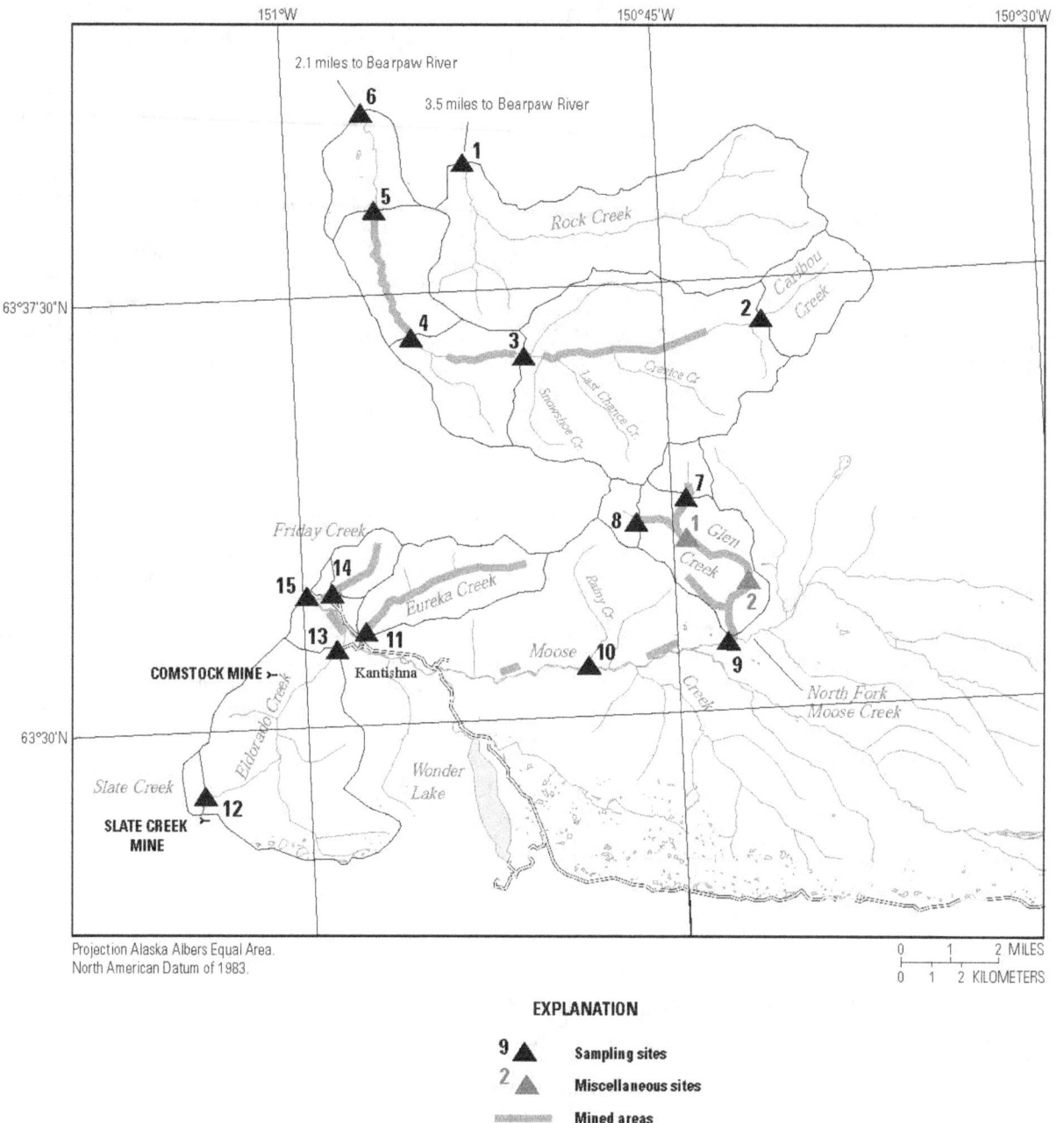

Figure 5.   Location of monitoring sites and mined areas in Kantishna Hills, Alaska.

Table 1. Water-quality data-collection sites in Kantishna Hills, Alaska.

[**Site No.:** Site locations are shown in figure 2. **Biology:** refers to macroinvertebrates and algae. **Abbreviations:** USGS, U.S. Geological Survey; mi^2, square mile; nd, not determined; –, no sample collected]

Site No.	USGS station No.	Latitude/longitude	Station name	Drainage area (mi^2)	Sampling dates			Continuous monitoring dates
					Biology	Water-quality	Fish	
1	15547510	63°40'01" 150°52'25"	Rock Creek 4.0 miles above Bearpaw River near Kantishna, Alaska	19.2	2008–09	2008–11	–	2009
2	15547520	63°37'00" 150°40'30"	Caribou Creek 3.4 miles above Crevice Creek near Kantishna, Alaska	3.8	2008–09	2008–10	–	2009
3	633631150501500	63°36'31" 150°50'15"	Caribou Creek 0.3 miles below Snowshoe Creek near Kantishna, Alaska	21.6	–	2008	–	–
4	633656150545800	63°36'56" 150°54'58"	Caribou Creek 7.9 miles above Bearpaw River near Kantishna, Alaska	26.1	–	2008	–	–
5	633914150561800	63°39'14" 150°56'18"	Caribou Creek 4.8 miles above Bearpaw River near Kantishna, Alaska	31.0	–	2008	–	–
6	15547530	63°40'56" 150°56'42"	Caribou Creek 2.6 miles above Bearpaw River near Kantishna, Alaska	34.2	2008–09	2008–11	2009	2009
7	633352150435500	63°33'52" 150°43'55"	East Fork Glen Creek 0.8 miles above mouth near Kantishna, Alaska	1.1	–	2008	–	–
8	633328150455900	63°33'28" 150°45'59"	West Fork Glen Creek 0.8 miles above mouth near Kantishna, Alaska	0.9	–	2008	–	–
9	15547560	63°31'16" 150°42'26"	Glen Creek 150 feet above mouth near Kantishna, Alaska	5.0	2008–09	2008–11	–	2009
10	633057150481100	63°30'57" 150°48'11"	Moose Creek 100 feet above Rainy Creek near Kantishna, Alaska	92.7	–	2008	–	–
11	15547570	63°31'26" 150°58'29"	Eureka Creek near Kantishna, Alaska	5.0	2008–09	2008–11	2009	–
12	15547575	63°28'52" 151°04'25"	Slate Creek 2.1 miles above Eldorado Creek near Kantishna, Alaska	0.5	2008–09	2008–11	–	–
13	15547580	63°31'26" 150°58'29"	Eldorado Creek 0.5 miles above mouth near Kantishna, Alaska	12.5	–	2008, 2010	–	–
14	15547590	63°32'26" 150°58'36"	Friday Creek near Kantishna, Alaska	1.5	2008–09	2008–09	–	–
15	15547600	63°32'24" 150°59'40"	Moose Creek 800 feet below Friday Creek near Kantishna, Alaska	152	–	2008–11	–	–
Miscellaneous sites								
1	633310150442000	63°33'10" 150°44'20"	Miscellaneous site 1 on Glen Creek– downstream of 2009 reclamation work	nd	–	–	–	2009
2	633228150414000	63°32'28" 150°41'40"	Miscellaneous site 2 on Glen Creek– downstream of 2009 reclamation work	nd	–	–	–	2009

Rock Creek (site 1)

At the monitoring site (fig. 6), Rock Creek drains 19.2 mi^2 and serves as a reference site. Rock Creek begins as several small forks on the main divide of the Kantishna Hills just northwest of Kankone Peak. The forks converge into the main stem that flows west and then north in an arc for about 12 mi before entering the Bearpaw River. Spruce and birch forests cover most of the drainage and willows are the predominant riparian vegetation type. Past mining is not evident in the Rock Creek watershed. However, at least some prospecting likely was done in the early 1900s, particularly during the Kantishna stampede in 1905.

Caribou Creek (sites 2–6)

At the most downstream monitoring site (site 6; figs. 2 and 5), Caribou Creek drains 34.2 mi^2, with elevations from 885 ft at the mouth to 4,990 ft near Kankone Peak. Caribou Creek flows west then north before joining the Bearpaw River (total length about 17 mi). The bedrock geology of the Caribou Creek watershed is composed of faulted and folded quartzite and hornblende schist of the Birch Creek formation. Thin (10 ft) gravel deposits upstream give way downstream to thicker deposits of subangular to subrounded clasts of polymetamorphic schist and quartz (Bundtzen, 1981).

Placer mining began in Caribou Creek in 1905. A large hydraulic mining operation worked on upper Caribou Creek in 1922 and a large dragline dredge also worked three full seasons, beginning in 1939, at several locations on upper Caribou Creek. In 1982, two small suction dredges operated on upper Caribou Creek. Two placer mining operations using heavy equipment began work in 1983 near the mouth of Last Chance creek and about 4 mi upstream of the mouth of Caribou Creek.

Caribou Creek can be divided into upper Caribou Creek and lower Caribou Creek. Upper Caribou Creek (upstream of site 3) includes the tributaries from the south (Crevice, Last Chance, and Snowshoe Creeks) and flows west-southwest and west for about 6 mi then turns north for about 3 mi. Initially, upper Caribou Creek flows through a fairly steep canyon, dropping about 500 ft in 1.5 mi before leveling out. The valley is fairly narrow, but at most places there is at least a 100 ft width of valley-bottom alluvium. Gravel thickness gradually increases downstream from a minimum of about 2 ft to about 12–14 ft. The two mined areas in this stretch of upper Caribou Creek were (1) upstream of the confluence of Crevice Creek and Caribou Creek (fig. 7), and (2) on Caribou Creek between the tributaries of Crevice Creek and Last Chance Creek (fig. 5). Lower Caribou Creek has a gradient of about 150 ft/mi and flows through a broad alluvial valley cut into bedrock of the Birch Creek Schist (fig. 8).

Figure 6.   Sampling section at site 1 at Rock Creek (looking downstream), Kantishna Hills area, Alaska, August 23, 2008. Channel width is 24 feet, average velocity is 2.6 feet per second, and flow is 66 cubic feet per second. Photograph taken by Dan Long, U.S. Geological Survey.

Figure 7.   Revegetation of mining areas near site 2 at Upper Caribou Creek (looking downstream), Kantishna Hills area, Alaska, June 12, 2008. Flow was 17.2 cubic feet per second. Area on right was mined in the 1980s and is slowly becoming revegetated. Photograph taken by Dan Long, U.S. Geological Survey.

Figure 8.   Sampling site 6 at Caribou Creek (looking downstream), Kantishna Hills area, Alaska, August 23, 2008. Channel width is 30 feet, average velocity is 2.6 feet per second, and flow is 106 cubic feet per second. Photograph taken by Dan Long, U.S. Geological Survey.

Glen Creek (sites 7–9)

The Glen Creek watershed drains 5.0 mi². Glen Creek begins as two forks, each about 1 mi long, and flows south from their confluence for approximately 3 mi before entering North Fork Moose Creek. Nearly the entire length of Glen Creek, upstream and downstream of the confluence of the east and west forks has been altered in conjunction with placer mining (fig. 5). Placer mining began on Glen Creek in 1906 and involved hand work. Little or no mining activities occurred from 1945 to 1961. Since the late 1960s, heavy equipment was used and resulted in extreme alteration of the lower 1.5 mi of Glen Creek. The entire width of the valley floor in the lower stretch has been cleared of vegetation and mined, and now consists of old tailing piles and filled settling ponds.

The bedrock geology of the Glen Creek watershed is faulted and folded quartzite and hornblende schist of the Birch Creek formation. The study area on lower Glen Creek was covered in the middle Wisconsin Glaciation with glacial ice from the Alaska Range, and gravel and rocks deposited by the glacier are mixed with bedrock material in the alluvial gravels. Trees are confined to favorable sites on alluvial terraces and south-facing slopes. Tall shrubs dominate riparian vegetation on the floodplains and younger terraces, and low shrubs and herbs form the tundra vegetation on colder, more exposed sites. Non-mined areas are dominated by feltleaf willow (10–13 ft tall) mixed with varying amounts of American Greenleaf alder (3–6 ft tall).

Eureka Creek (site 11)

Eureka Creek drains an area of 5.0 mi² (fig. 5). Eureka Creek, consisting primarily of the Birch Creek Schist, was the richest stream mined in the Kantishna district and probably the second in total production. Total production probably exceeded 10,000 ounces of gold, of which about three-fourths was recovered in 1905–06 (Meyer and Kavanagh, 1983). The lower 3 mi of the stream has been heavily mined; probably very little, if any, of the original stream channel remains intact. In most places the entire width of the valley floor has been mined (fig. 9).

Slate Creek/Eldorado Creek (sites 12–13)

Eldorado Creek flows north and northeast on the south flank of the Kantishna Hills and drains 12.5 mi². Slate Creek is a tributary to Eldorado Creek and at the monitoring site drains 0.5 mi². One of the main antimony deposits in the Kantishna Hills is in the Slate Creek watershed and occurs as a quartzite unit of the Birch Creek Schist. The mine produced about 679 tons of hand-picked antimony ore; production from about 1916 to 1945 (end of World War II) totaled about 325 tons of ore averaging 49 percent antimony. From World War II to 1983, the mine produced about 354 tons of ore that averaged about 60 percent antimony. About 22 tons of antimony ore was produced in 1983.

In addition to the Slate Creek Mine, other mining occurred in the Eldorado Creek watershed. The Comstock Mine (also known as the Bonnell or Neversweat Mine) is located approximately 3 mi upstream of its confluence with Moose Creek. Exact production dates are unknown but the mine seems to have been inactive since the 1960s. Mining also occurred along Eldorado Creek near its mouth at Moose Creek. Mining began around 1916 and was abandoned for many years until 1975. In 1982, two placer mining operations using backhoes worked a large area of the Eldorado-Moose Creek floodplain. Water was drawn from the Eldorado and Moose Creeks for wash plant operations, but all the effluent was discharged into a series of large settling ponds constructed parallel to Moose Creek. Eldorado Creek stream channels were cut and moved several times in the process of mining different areas.

Friday Creek (site 14)

At the monitoring site, Friday Creek drains an area of 1.5 mi² (fig. 5). Friday Creek begins on the west side of Wickersham Dome and flows west for about 1.7 mi before entering Moose Creek. The watershed is underlain mainly by metafelsite and chlorite phyllite of the lower Spruce Creek sequence. The streambed is fairly steep, about 400 ft/mi, above an inflection point near its junction with Moose Creek. Two main types of placer deposit have been exploited; alluvial placers along Friday Creek, and complex fan and terrace gravel deposits where Friday Creek enters Moose Creek. The amount of gold recovered from Friday Creek is uncertain. It is estimated that the amount of gold recovered before 1980 was less than 1,000 ounces and in 1982–83 more than 4,000 ounces (Meyer and Kavanagh, 1983). Total production from Friday Creek probably is 5,000–10,000 ounces of gold. At least two placer operations that worked down to bedrock were active in 1982. The largest operation was near the road crossing and the other was approximately 0.5 mi upstream. An undetermined number of settling ponds were constructed near the road crossing and about 0.3 mi upstream, but the ponds filled in rapidly. By the end of the summer of 1982 considerable alterations had been made to about 40 percent of the entire stream length. Alterations included clearing of riparian vegetation, overburden removal, relocation of the stream channel, channelization, and construction of settling ponds (fig. 10).

Figure 9. Sampling site 11 at Eureka Creek (looking upstream), Kantishna Hills area, Alaska, August 21, 2008. Channel width is 12 feet, average velocity is 2.4 feet per second, and flow is 17 cubic feet per second. Photograph taken by Dan Long, U.S. Geological Survey.

Figure 10. Sampling site 14 at Friday Creek (looking upstream), Kantishna Hills area, Alaska, June 13, 2008. Channel width is 9.5 feet, average velocity is 1 foot per second, and flow is 1.3 cubic feet per second. Area on left is old mine tailings. Photograph taken by Dan Long, U.S. Geological Survey.

Moose Creek (sites 10 and 15)

Moose Creek drains an area of 152 mi^2. In addition to the fans or mouths of Eldorado, Eureka, Friday, and Glen Creeks, two other reaches of Moose Creek were mined (fig. 5). Mining began in 1905 (Meyer and Kavanagh, 1983) and in 1922, one of the largest operations on Moose Creek upstream of Eureka Creek reportedly cleared about 18,000 yds^3 of gravel. Little or no placer mining activities occurred during 1945–60. From 1960 to 1966, a prospecting venture operated on the north side of Moose Creek, 1 mi below the mouth of Rainy Creek. From 1969 to 1982, there were three placer mining operations along Moose Creek. Two operations were near Kantishna (figs. 11 and 12) and the other operation was 0.5 mi downstream from the confluence with the North Fork of Moose Creek. The operations near Kantishna worked the gravels on the west side of Moose Creek and discharged wash water into a series of settling ponds incorporating water from Eldorado Creek. The operation near the North Fork of Moose Creek worked a broad, flat-gravel floodplain adjacent to the river and discharged sluice water into a large settling pond.

Figure 11. Active mining in the Moose Creek valley at Kantishna, Alaska, during the 1970s. View is looking due south toward Mt. McKinley (in background) from Friday Creek. Note Eldorado Creek. Photograph courtesy of Denali National Park and Preserve.

Figure 12.   Moose Creek valley at Kantishna, Alaska, June 12, 2010. View is looking southeast near the Kantishna airstrip. Note that much of the valley is revegetated. Photograph taken by Tim Brabets, U.S. Geological Survey.

Past and Present Reclamation Efforts

Restoration activities at DENA fall into two categories: (1) removing hazardous materials, contaminated soils, and abandoned equipment and (2) reconstructing floodplain and stream channel structures to stabilize stream banks from erosion and revegetating sites. For the second category, procedures have been developed based on a long-term stream restoration and monitoring project, started in lower Glen Creek in 1991 (Karle and Densmore, 1994). Initially, 100-year design flood flows were estimated from regional multiple-regression techniques, and slope and sinuosity determinations were made by regional comparisons to other Kantishna streams. Shear stress equations were applied to determine bed and bank stability. However, channel monitoring showed substantial changes from smaller annual flows (Karle and Densmore, 1994), due to the unlimited sediment supply from tailing piles, incised channel conditions, and denuded floodplains. Thus, newer design guidelines place greater emphasis on addressing excess sediment supply in the watershed.

Revegetation is an essential component of restoration. The root system anchors the substrate, and above-ground stems decrease water velocity, catch organic debris, and promote sediment deposition. Similar to floodplain reconstruction, initial estimates of revegetation growth were slower than expected in the Kantishna Hills area because of the lack of soil and low nutrient levels in the mine tailings. Thus, newer design guidelines which use native plant materials have been compiled (Densmore and others, 2000), and are followed for revegetation efforts.

Caribou Creek

Caribou Creek has the most extensive placer mining related effect of any drainage in DENA. As previously noted, Caribou Creek was dredged in the 1930s and then aggressively reworked with heavy equipment over most of its length in the 1970s and 1980s (fig. 13). In 2001, two large bulldozers were brought to the site to flatten tailing piles, remove an abandoned airstrip, and route a new sinuous channel through the stream channel. In 2002, the project continued with a single bulldozer to conduct additional earthwork, create new channel sections, and install coconut fiber coir bio-logs along the outside bends of several channel meanders.

To reduce sediment sources at upper Caribou Creek during high flows, a 125 ft section of bank (which was the side slope of a remnant tailings pile) was moved away from the stream in 2010, widening the floodplain by more than 15 ft. This floodplain bench was established 2 ft above the adjacent thalweg and provides relief for higher flows in a formerly constricted area of the stream. Another section of Caribou Creek was straightened (radius of curvature was too small), moved to the center of the valley, and shortened by 500 ft. The design implemented in 2010 was to use the coconut fiber coir bio-logs that were used during reclamation efforts in 2002. However, the bio-logs were missing or unusable and on-site modifications were necessary. Crews used a combination of stabilizing existing logs in place, rewrapping staged logs on-site with new jute material, or installing new logs. At completion, all locations prescribed for bank stabilization received treatment similar or equal to the design, including the construction of four rock weirs to direct flows away from the unvegetated bank during high flows.

Glen Creek

Beginning in 1991, most of the work along the east and west forks of Glen Creek involved recontouring artificially raised floodplains to a lower elevation and leaving the existing channel undisturbed except for minor bank modifications. This effort served to reconnect the channel to the floodplain by restoring natural floodplain processes and bringing the desired riparian or floodplain surface closer to the water table. In the lower study reach, a 500 ft long section was moved from an incised location against the valley wall to the center of the valley. Excavated gravels were used in some areas to fill in settling ponds, old channel beds, and other mining-related topographic features. Excess gravels were contoured into the valley slope at the floodplain edge. In 1992, several bioengineering techniques for floodplain stabilization and revegetation were tested, including brush bars, willow cuttings, and stream bank brush layering.

Figure 13.   Mining near site 2 at Upper Caribou Creek (looking downstream), Kantishna Hills area, Alaska, in the mid-1980s. Berms were constructed along Caribou Creek and the adjacent hillside was mined for gold. Photograph courtesy of Denali National Park and Preserve.

In 2009, reclamation work included removal of mining debris and equipment, floodplain contouring, channel construction, bank treatments, installation of instream controls, tailings pile earthwork, and revegetation (figs. 14 and 15). Abandoned equipment and mining debris, which ranged from screening plants to school buses, were removed from Glen Creek and either transferred to Kantishna property owners for reuse (for example, reusable buildings), recycled (for example, scrap metal), staged for burning (for example, wood debris) or disposed. The range and volume of debris removed required substantial planning and time. The size, steepness, and location of a particular tailings pile precluded complete removal from the floodplain. Instead, gabion baskets and Reno mattresses were installed at the toe of the slope to prevent lateral erosion of the creek into the tailings pile.

Figure 14. Conditions before reclamation of Upper Glen Creek, Kantishna Hills area, Alaska, June 2, 2009. View is looking upstream at the confluence of the East Fork Glen Creek (right) and West Fork Glen Creek (left). Photograph taken by Tim Brabets, U.S. Geological Survey.

Figure 15. Conditions after reclamation of Upper Glen Creek, Kantishna Hills area, Alaska, September 10, 2009. View is looking upstream at the mouth of East Fork Glen Creek (right) and West Fork Glen Creek (left). Reclamation consisted primarily of removal of old mining equipment, leveling of tailings, and hydroseeding. Photograph taken by Tim Brabets, U.S. Geological Survey.

Slate Creek

In 1997, upland restoration, designed to improve drainage from the exposed ore body at the antimony mine site, was accomplished by recontouring the mine area to direct drainage away from the stream channel. Several thousand cubic yards of tailings were removed from the floodplain and hillslope and returned to the antimony pit to cover up acid-generating exposures. Tailing piles were moved inside the mine pit and leveled. Geo-textiles were used to intercept and modify groundwater flows. An old settling pond and trench were drained of water and filled with material.

In 2010, excavation of an extended tailings pile at Slate Creek was one of the significant challenges. The tailings comprised of fine silt, clay, and gravels were saturated from hillside seeps and precipitation. Excavation of the fine material, in a saturated condition, resulted in occasional liquefaction of the tailing pile. Excavating this extended tailings pile removed a substantial volume of loose and unstable substrate from the floodplain that was contributing to accelerated erosion and sedimentation, increased water turbidity, and water-quality degradation to Slate, Eldorado, and to some extent, Moose Creek. The new slope, where excavated material was removed, was constructed to match the existing natural valley side slopes and was treated with biodegradable erosion blankets.

A new 400 ft channel was constructed to redirect streamflows around the main pit area of the former mine (figs. 16 and 17). Fabric-encapsulated soil lifts (FESLs) were used to construct the new stream banks. Additionally, willow cuttings were inserted into the lifts for revegetation and long-term stability. Rock was placed in the channel to provide roughness in the steep channel (7–10 percent slope). Revegetation consisted of scarifying and grading tailings piles and roadbeds, tracking slopes, transplanting vegetation, installing willow cuttings or erosion blankets, seeding with native plants and temporary erosion controls using ryegrass, hydroseeding, liming, and fertilizing at prescribed rates. Nearly 4,000 willow cuttings were installed.

Figure 16.   Conditions before reclamation at Slate Creek, Kantishna Hills area, Alaska, June 1, 2010. View is from the headwaters of the stream looking downstream; the antimony mine is in the foreground. Stream entering from the left foreground drains natural area. Photograph taken by Tim Brabets, U.S. Geological Survey.

Figure 17. Conditions after reclamation at Slate Creek, Kantishna Hills area, Alaska, September 2010. Note that the stream from the left has been rerouted around the former mine, ground has been recontoured, and erosion blankets have been installed along steep slopes. Photograph courtesy of Denali National Park and Preserve.

Eureka Creek

In 1999, reclamation consisted of four phases: (1) mining equipment relocation, (2) hazardous waste assessment and removal, (3) debris removal, and (4) stream channel and floodplain reconstruction. A topographic survey determined the existing channel geometry and valley topography. Using a combination of geomorphology and standard open-channel hydraulic equations, a new channel and floodplain were designed that would allow the channel to carry as much as a bankfull discharge before overtopping and allow the floodplain to carry as much as a 1 percent annual exceedance probability flood. After completion of the designed floodplain, about 500 ft of Eureka Creek received bank stabilization treatments, which involved the installation of double or triple rows of 1-ft coir bio-logs. These logs were staked and anchored at the channel edge along the outside of the newly created bends.

Moose Creek

This short-term reclamation project began in May 2000 and was completed by September 2000. Ten 55-gallon barrels, in poor condition, were removed. A large steel building located on upper Moose Creek near the mouth of Glen Creek was dismantled and removed from DENA. Surrounding debris, including four 55-gallon barrels, and other miscellaneous steel, also was removed and properly disposed. Many other pieces of mining-related heavy equipment within the Moose Creek floodplain were removed.

Friday and Eldorado Creeks

At Friday Creek, the area near the mouth was converted into a camp for use by National Park Service personnel during the summers of 1987 and 1988. Work during the summers of 1987 and 1988 also included removal of hazardous material and flattening tailings piles. In 2010, mining debris was removed from near the mouth of Eldorado Creek in the Moose Creek floodplain.

Methods of Data Collection and Analysis

Data Collection

For this study, characterization of the water quality conditions in the Kantishna Hills included 15 stream sites for water quality, 7 sites for benthic communities, and 2 sites for fish tissue (fig. 5, table 1). The stream sites sampled for water quality were located on eight streams and the sites sampled for benthic communities were located on six streams. Water-quality data were collected during the summers of 2008–11 and biological data were collected during the summers of 2008–09.

Water samples collected from the streams were analyzed for dissolved major ions, total dissolved solids, dissolved trace elements, suspended sediment, dissolved organic carbon, and phytoplankton (table 2). The sample collection and processing equipment was made from Teflon™, glass, or stainless steel to prevent sample contamination and to minimize analyte losses through adsorption. All sampling equipment was cleaned prior to use with a non-phosphate laboratory detergent and deionized water, followed by a native-water rinse prior to sample collection. Depth-integrated water samples were collected across the streams using the equal-width-increment method or the centroid-of-flow method (U.S. Geological Survey, variously dated) and processed in the field using methods and equipment described by Shelton (1994). Samples to be analyzed for dissolved constituents were filtered through 0.45 μm capsule filters. Water samples were sent to the U.S. Geological Survey (USGS) National Water-Quality Laboratory (NWQL) for analysis using standard USGS analytical methods (Fishman and Friedman, 1989; Patton and Truitt, 1992; Fishman, 1993).

A Yellow Springs Instrument™ (YSI) multiparameter sonde was used to measure field parameters (water temperature, specific conductance, pH, dissolved-oxygen concentration, and turbidity) at the time of sampling. YSI sondes were installed at six sites in 2009 to continuously collect turbidity, water temperature, and specific conductance during the open-water season. Instruments were checked for calibration at each site visit and data were processed using methods outlined by Wagner and others (2006). Streamflow discharge measurements also were made at the time of sampling using methods of Turnipseed and Sauer (2010).

Slimy sculpin, *Cottus cognatus*, was selected as the most appropriate species for tissue analysis. This species is non-migratory and a bottom-feeding omnivore, which are characteristics necessary for interpretation of results (Crawford and Luoma, 1993). Crawford and Luoma (1993) recommend analysis of whole fish livers for trace elements. Because adult slimy sculpin are small, the whole fish was used for analysis of trace elements.

Fish were collected using a baited fish trap. Slimy sculpin captured in the fish trap were placed in a bucket of stream water until sufficient numbers had been collected to constitute a sample (30 g). Each fish in the composite sample was weighed and measured. Fish used for trace element samples were double bagged in plastic. Composited samples were immediately placed on wet ice and then into a portable freezer. Whole sculpin were analyzed for trace elements (table 2) at the NWQL using methods described by Hoffman (1996).

Streambed sediments were sampled from several depositional areas at 14 sites and analyzed for trace elements, forms of carbon, and major elements (table 2). Streambed sediment samples were collected from depositional areas accessible by wading in the selected stream reaches. Typically, these depositional areas were along channel margins or bar formations. The surficial 2 in. of sediment was collected from several areas and composited to form a large sample volume from which material was removed and sieved using a 0.063-mm nylon mesh (Shelton and Capel, 1994). Because the concentration of trace elements on streambed materials is strongly affected by the particle-size distribution of the sample, only that portion of the sample finer than 63-μm was analyzed. Stream water was used for sieving the trace-element sample through a 63-μm mesh. Water included in the trace-element sample was decanted after very fine-grained sediments had settled (about 24 hours). The streambed sediment samples were analyzed by the USGS NWQL following a total digestion procedure described by Arbogast (1990).

Benthic macroinvertebrates were collected at seven stream sites in 2008–09 (table 1). Richest Targeted Habitat (RTH) and Qualitative Multi-Habitat (QMH) samples were collected and processed using protocols established by the National Water-Quality Assessment (NAWQA) program (Cuffney and others, 1993) and were identified by EcoAnalysts, Inc., Moscow, Idaho. RTH samples are semi-quantitative; providing taxa abundance and richness measures by sampling from a delineated area to a predetermined depth of bed sediment, and carefully removing macroinvertebrates that were subsequently washed downstream into a fine (425-μm mesh) net. Five sites within the reach in stream riffles were selected in advance as collection sites. Samples from the five sites were combined into a composite sample for processing, identification, and enumeration. QMH samples collected with a standard D-frame net (210-μm mesh net) are a measure of the presence or absence of macroinvertebrates in available microhabitats to give a broad accounting of organisms within the stream reach.

Table 2.   Analyses made on water samples and streambed sediments collected, Kantishna Hills, Alaska, 2008–11.

[**Abbreviations:** °C, degrees Celsius; mg/L, milligram per liter; FNU, Formazin Nephelometric Unit; ft³/s, cubic foot per second; µg/g, microgram per gram; µg/L, microgram per liter; µS/m at 25°C, microsiemens per centimeter at 25 degrees Celsius]

Parameter or constituent	Reporting level
Physical field parameters	
Streamflow (ft³/s)	0.01
Dissolved oxygen (mg/L)	0.1
pH (units)	0.1
Specific conductance (µS/cm at 25°C)	5
Water temperature (°C)	0.1
Turbidity (FNU)	2
Chemical constituents in water (mg/L)	
Alkalinity	1
Bicarbonate	1
Calcium	0.022
Chloride	0.06
Total dissolved solids	12
Flouride	0.04
Magnesium	0.008
Potassium	0.022
Silica	0.029
Sodium	0.06
Sulfate	0.09
Trace elements in water and fish tissue (µg/L)	
Aluminum	1.7
Antimony	0.027
Arsenic	0.022
Barium	0.07
Beryllium	0.01
Boron	6
Cadmium	0.04
Chromium	0.12
Cobalt	0.02
Copper	1
Iron	3.2
Lead	0.08
Lithium	1
Manganese	0.13
Molybdenum	0.2
Nickel	0.09
Selenium	0.03
Silver	0.1
Strontium	0.2
Thallium	0.04
Uranium	0.004
Vanadium	0.08
Zinc	1.4
Suspended sediment in water (mg/L)	
Concentration	1

Parameter or constituent	Reporting level
Organics in water	
Dissolved organic carbon (mg/L)	0.15
Pheophytin phytoplankton (µg/L)	0.1
Chlorophyll-a phytoplankton (µg/L)	0.1
Trace elements in streambed sediments (µg/g)	
Aluminum	850
Antimony	0.04
Arsenic	1
Barium	0.2
Berylium	0.03
Bismuth	0.06
Cadmium	0.007
Cerium	0.1
Cesium	0.05
Chromium	0.5
Cobalt	0.02
Copper	2
Gallium	0.02
Iron	5
Lanthanum	0.05
Lead	0.4
Lithium	0.3
Manganese	0.7
Mercury	0.01
Molybdenum	0.05
Nickel	0.3
Niobium	1
Scandium	0.04
Selenium	0.1
Silver	2
Strontium	0.8
Thallium	0.08
Thorium	0.1
Titanium	40
Uranium	0.02
Vanadium	0.2
Yittrium	0.05
Zinc	3
Major elements and forms of carbon in streambed sediments	
Inorganic carbon (percent)	0.01
Organic carbon (percent)	0.01
Total carbon (percent)	0.01
Calcium (µg/g)	100
Magnesium (µg/g)	6
Phosphorus (µg/g)	5
Potassium (µg/g)	20
Sodium (µg/g)	20
Sulfur (percent)	0.05

In 2008–09, RTH and QMH periphytic algae samples were collected according to protocols outlined by Porter and others (1993) from seven sites (table 1). An assessment of the algal community was done to identify the species of algae present at multiple habitats within each stream reach. Algae samples were collected in habitats similar to the RTH and QMH benthic macroinvertebrate samples (depositional zones, woody debris, and rocks) and were analyzed by the Philadelphia Academy of Natural Science, Philadelphia, Pennsylvania.

Quality Assurance

An important part of the study was the calibration of field equipment and collection and analysis of quality-control samples. Each day the YSI multiparameter sonde was used, the probes (dissolved oxygen, pH, specific conductance, and turbidity) were calibrated to known standards and a log was kept of all calibrations. Quality control samples consisted of field blank and concurrent replicate sample analyses. The primary purpose of a field blank sample is to determine if the sampling equipment has been adequately cleaned, and the primary purpose of a concurrent replicate sample is to determine the variability (if any) introduced from collection, processing, and laboratory handling of the sample. For each year of data collection, 2008–11, one blank sample and one replicate sample were collected during the data collection period, late May to mid-September. Both organic-free and inorganic-free water were used for the blank sample. Results of the blank samples showed no or minimal contamination and results of the replicate sampling showed the same result as the environmental sample. All calibration and quality-assurance data are available for inspection at the USGS Anchorage office.

Quality-assurance and quality-control results for the macroinvertebrate and algae samples for the Kantishna Hills study were minimal. For the macroinvertebrates, EcoAnalysts returned a reference collection of all macroinvertebrates that were identified, including slide mounts of members of the family Chironomidae (midges) collected both years (2008–09). The reference collection was checked for accuracy, and the identifications were determined to be satisfactory. Although there was no on-site quality assurance/quality control for algae collection, the quality assurance/quality control protocols are very detailed at the Patrick Center for Environmental Research, Philadelphia Academy of Natural Science, where the samples were analyzed.

Data Analysis

Data analysis started after the data were collected, checked, and compiled. The concentrations of various water-quality constituents in the sampled streams were compared to published concentrations for drinking water and fish survival. Trace-element concentrations in streambed sediments were compared with those collected by the USGS on streams located on the south side of DENA, and with guidelines established by the Canadian Council of Ministers of the Environment (CCME; 1999). Water-quality data also were compared with data collected from the same streams in previous studies (West, 1982; Meyer and Kavanagh, 1983; West and Deschu, 1984; Deschu, 1985a, 1985b, 1986; Deschu and Kavanagh, 1986; Edwards and Tranel, 1997) to determine if water quality had changed. Macroinvertebrate and algae data were compared with similar data collected from other sites in Alaska using the NAWQA sampling and analysis protocols.

Converting RTH and QMH macroinvertebrate data to metrics used for comparing or contrasting sites was accomplished using the Invertebrate Data Analysis System, version 5.0.0, (IDAS) (Cuffney and Brightbill, 2011). The IDAS application is designed to parse, organize, and calculate metrics related to macroinvertebrate data collected as part of the NAWQA protocols. The software adds accuracy and precision to data interpretation by tracking and documenting settings used in metric generation, thereby yielding easily reproducible and comparable results. The options for data processing were set for the default, with the exception of the combination of parent taxa with children. Applicable non-density dependent metrics combined RTH and QMH data for the calculation of qualitative metrics.

Macroinvertebrate metrics that have been identified as sensitive to the difference between undeveloped land use and other land use types were calculated for the Kantishna Hills sites and other Alaska sites using IDAS. The National Invertebrate Community Ranking Index (NICRI), was used to rank each of the samples collected. Each metric is ranked according to a percent rank function that orders each metric value for a sample relative to other values for the metric. Percent ranks of metrics that exhibit a positive response to disturbance (abundance of tolerant taxa, for example) were subtracted from one (1-percent rank) to maintain a consistent, standardized order of values within and between metrics. The rankings of each metric were then added together to determine the final summary order. The Kantishna Hills samples were ranked relative to each other to see whether the samples follow expected trends based on past mining activities at each site.

Preparation of the RTH and QMH algae data was accomplished using the Algal Data Analysis System (ADAS) (Cuffney and Brightbill, U.S. Geological Survey, written commun., 2010). ADAS was designed to parse, organize, and calculate metrics related to algal data. Samples were kept separate while resolving ambiguities and parents were distributed among children for both RTH and QMH data. Taxa richness for each sample was calculated using both RTH and QMH data; abundance, diversity values, and multivariate analyses were generated using only the RTH data. Abundance data are reported for density (number of individuals) and biovolume (algal biomass). The advantage of reporting abundance as biovolume in addition to density is that biovolume accounts for cell-size variation between taxa.

Macroinvertebrate and algae data were further analyzed using the ordination technique, non-metric multidimensional scaling (NMDS) (Venables and Ripley, 2002; Oksanen and others, 2010) using the R software (R Development Core Team, 2010) and specifically the vegan (Oksanen and others, 2010) and Modern Applied Statistics with S (MASS) (Venables and Ripley, 2002) packages. Macroinvertebrate presence or absence data were standardized to the lowest identifiable taxonomic level for each sample to generate a Bray-Curtis dissimilarity (distance) input matrix (Bray and Curtis, 1957). NMDS plots were then generated from the two distance matrices for the Kantishna Hills samples, and the Kantishna Hills samples combined with other Alaska macroinvertebrate or algae samples.

Water-Quality Characteristics

Water-quality data collected at streams located in Kantishna Hills in 2008–11, included (1) measurements of streamflow, specific conductance, pH, water temperature, dissolved-oxygen concentration, and turbidity, (2) collection and analysis of water samples for major inorganic ions, dissolved organic carbon (DOC), trace elements, and suspended sediment, and (3) collection and analysis of streambed sediments for trace elements. The analysis of the data collected in this study together with water-quality data collected in the 1980s and 1990s provided a good overview of the current water-quality conditions of the streams in the Kantishna Hills and whether water quality has changed.

Streamflow

Typical of interior Alaska, most of the annual streamflow in the Kantishna Hills area is from mid-May to mid-October. For non-glacier streams such as those in Kantishna Hills, above average streamflow occurs in late May or early June due to snowmelt. From early June through September streamflow remains at average conditions unless rainfall occurs. In October, as air temperatures cool and rainfall ceases, streamflow gradually becomes lower and is solely a function of groundwater discharge into the stream.

During the study period the highest streamflows, as expected, were measured at Moose Creek (site 15), the largest watershed (152 mi²), and ranged from 92 to 320 ft³/s (table 3). Also expected, the lowest streamflows were measured at Slate Creek (site 12), the smallest watershed (0.5 mi²), and ranged from 0.3 to 2.1 ft³/s. Streamflows on Caribou Creek (sites 2–6) showed an increase in the downstream direction, indicating that Caribou Creek is a gaining stream. Streamflows on Glen Creek (sites 7–9) also indicated a gaining stream.

To compare streamflows between the unmined stream, Rock Creek, and mined streams, streamflows were converted to unit streamflows of cubic feet per second per square mile [(ft³/s)/mi²] (table 3). Boxplots, which provide visual summaries of the center of the data (median), the variation or spread (interquartile range), the skewness, and the presence of unusual values, were constructed for unit flow for all mined stream sites and for Rock Creek (fig. 18). Comparison of the boxplots showed nearly identical values for the median, 25th and 75th percentile for all unit streamflows at mined streams and unit streamflows at Rock Creek, indicating that during the period of study, runoff characteristics were similar between mined and unmined streams.

Table 3.   Physical field parameters, suspended sediment, and dissolved organic carbon measured at surface water sites in Kantishna Hills, Alaska, 2008–11.

[Number below constituent in parentheses is used by U.S. Environmental Protection Agency and U.S. Geological Survey to identify parameters in computer databases. **Abbreviations:** ft^3/s, cubic foot per second; (ft^3/s)/mi^2, cubic foot per second per square mile; mg/L, milligram per liter; µS/cm at 25°C, microsiemens per centimeter at 25 degrees Celsius; °C, degrees Celsius; FNU, Formazin Nephelometric Unit; mi^2, square mile; <, less than; –, not measured; E, estimated]

Site No.	Date	Time	Streamflow (ft^3/s) (00061)	Unit streamflow [(ft^3/s)/mi^2]	Dissolved oxygen (mg/L) (00300)	pH (units) (00400)	Specific conductance (µS/cm at 25°C) (00095)	Water temperature (°C) (00010)	Turbidity (FNU) (63680)	Suspended sediment (mg/L) (80154)	Dissolved organic carbon (mg/L) (00681)
					Rock Creek (drainage area 19.2 mi^2)						
1	06-10-2008	1200	48.0	2.5	12.0	8.0	269	5.6	<2	2	4.2
	07-13-2008	1410	31.0	1.6	12.1	8.1	333	6.4	–	1	2.5
	08-23-2008	1000	66.0	3.4	13.1	8.0	318	2.9	<2	–	–
	09-10-2008	0955	36.0	1.9	12.6	8.2	364	3.1	<2	1	1.4
	06-03-2009	1420	27.0	1.4	10.3	8.2	303	9.6	<2	1	3.0
	07-28-2009	1620	15.0	0.8	10.8	8.0	381	11.5	<2	1	1.5
	08-12-2009	1425	13.8	0.7	11.9	8.2	411	10.0	<2	–	–
	09-10-2009	1000	25.5	1.3	13.8	8.1	411	3.9	<2	1	1.4
	06-01-2010	1320	13.8	0.7	9.3	8.4	357	8.1	<2	1	2.1
	07-13-2010	1355	33.8	1.8	11.7	8.1	322	7.3	4	1	2.9
	09-07-2010	1310	46.2	2.4	12.5	8.0	359	6.1	<2	<1	1.5
	06-07-2011	1420	36.1	1.9	12.9	8.0	281	6.5	<2	<1	3.2
	07-21-2011	1130	29.4	1.5	12.3	8.2	360	8.7	<2	<1	1.8
	09-06-2011	1245	55.5	2.9	11.9	8.3	348	6.0	<2	<1	2.0
					Caribou Creek (drainage area 3.8 mi^2)						
2	06-12-2008	1120	17.0	4.5	12.3	8.1	538	2.7	<2	1	–
	07-15-2008	1019	6.8	1.8	–	8.1	690	6.0	–	8	0.8
	08-24-2008	1120	14.0	3.7	12.7	8.2	594	3.8	12	–	–
	09-10-2008	1645	7.4	1.9	11.4	8.3	646	5.0	<2	5	0.6
	06-03-2009	1045	6.9	1.8	11.0	8.2	628	4.7	<2	1	1.1
	07-28-2009	1305	3.3	0.9	10.7	8.1	734	9.3	<2	1	0.6
	08-11-2009	1140	2.8	0.7	12.2	8.0	742	6.1	<2	–	–
	09-09-2009	1230	7.6	2.0	11.2	8.2	767	6.5	<2	2	0.5
	06-01-2010	1200	3.2	0.8	10.0	8.2	670	3.2	<2	<1	0.9
	07-13-2010	1120	7.3	1.9	11.4	8.1	658	6.4	<2	5	0.8
	09-07-2010	1223	12.1	3.2	12.0	7.8	661	5.7	<2	<1	E0.6
					Caribou Creek (drainage area 21.6 mi^2)						
3	06-12-2008	940	68	3.1	12.6	8.0	441	3.0	<2	1	2.6
	07-15-2008	1120	33	1.5	–	8.2	576	8.5	–	1	1.5
	09-10-2008	1530	46	2.1	11.6	8.2	604	5.8	<2	1	1.0
					Caribou Creek (drainage area 26.1 mi^2)						
4	06-10-2008	1645	70	2.7	11.8	8.1	411	6.1	<2	1	3.5
	07-15-2008	1240	38	1.5	–	7.9	514	9.5	–	1	1.9
	09-10-2008	1410	51	2.0	11.8	8.3	564	5.6	<2	1	1.2
					Caribou Creek (drainage area 31.0 mi^2)						
5	06-10-2008	1345	73	2.4	11.8	8.1	385	6.7	<2	2	4.2
	07-13-2008	1641	48	1.5	11.3	8.2	460	9.0	–	1	2.9
	09-10-2008	1300	51	1.6	11.9	8.3	541	5.3	<2	2	1.4

Table 3. Physical field parameters, suspended sediment, and dissolved organic carbon measured at surface water sites in Kantishna Hills, Alaska, 2008–2011.—Continued

[Number below constituent in parentheses is used by U.S. Environmental Protection Agency and U.S. Geological Survey to identify parameters in computer databases. **Abbreviations:** ft^3/s, cubic foot per second; (ft^3/s)/mi^2, cubic foot per second per square mile; mg/L, milligram per liter; µS/cm at 25°C, microsiemens per centimeter at 25 degrees Celsius; °C, degrees Celsius; FNU, Formazin Nephelometric Unit; mi^2, square mile; <, less than; –, not measured; E, estimated]

Site No.	Date	Time	Streamflow (ft^3/s) (00061)	Unit streamflow [(ft^3/s)/mi^2]	Dissolved oxygen (mg/L) (00300)	pH (units) (00400)	Specific conductance (µS/cm at 25°C) (00095)	Water temperature (°C) (00010)	Turbidity (FNU) (63680)	Suspended sediment (mg/L) (80154)	Dissolved organic carbon (mg/L) (00681)
					Caribou Creek (drainage area 34.2 mi^2)						
6	06-10-2008	1520	93	2.7	11.5	8.1	375	7.6	<2	1	4.2
	07-13-2008	1531	54	1.6	11.4	8.1	432	9.2	–	1	3.5
	08-23-2008	1430	106	3.1	11.4	8.2	458	8.5	<2	–	–
	09-10-2008	1150	56	1.6	12.1	8.3	527	5.2	<2	1	1.6
	06-03-2009	1230	47	1.4	10.5	8.2	410	8.8	<2	1	3.5
	07-28-2009	1450	24	0.7	10.2	8.1	534	14.2	<2	1	1.8
	08-11-2009	1500	21	0.6	11.4	8.2	564	12.5	<2	–	–
	09-09-2009	1730	43	1.3	11.1	8.1	573	9.3	<2	1	1.6
	06-01-2010	1420	28	0.8	9.1	8.4	472	10.0	<2	1	2.3
	07-13-2010	1240	54	1.6	11.2	8.1	432	9.6	<2	3	3.5
	09-07-2010	1400	78	2.3	12.2	8.2	501	7.7	<2	1	2.1
	06-07-2011	1240	55	1.6	12.7	8.0	378	7.2	<2	1	3.4
	07-21-2011	1230	41	1.2	11.6	8.2	469	11.7	<2	1	2.5
	09-06-2011	1140	79	2.3	11.9	8.2	477	6.3	<2	1	2.2
					East Fork Glen Creek (drainage area 1.1 mi^2)						
7	6-12-2008	1445	7.2	6.5	11.9	8.2	528	3.5	<2	1	1.2
	7-15-2008	1505	1.7	1.5	–	8.0	670	8.0	<2	1	0.8
	9-11-2008	0930	3.1	2.8	11.5	8.2	887	4.6	–	1	0.8
					West Fork Glen Creek (drainage area 0.9 mi^2)						
8	06-12-2008	1300	3.8	4.2	12.4	8.0	316	1.7	<2	1	2.6
	07-15-2008	1405	1.2	1.3	--	7.9	538	5.5	<2	1	0.9
	09-11-2008	1535	2.1	2.3	11.6	7.7	562	4.5	<2	1	0.7
					Glen Creek (drainage area 5.0 mi^2)						
9	06-11-2008	1140	14	2.8	11.5	8.2	430	5.2	<2	1	3.6
	07-15-2008	1050	6.5	1.3	–	8.0	567	11.0	<2	1	1.4
	08-22-2008	1050	19	3.8	11.3	8.4	578	4.9	<2	–	–
	09-09-2008	1020	11	2.2	12.0	8.2	653	3.5	<2	1	0.8
	06-02-2009	1635	5.5	1.1	9.5	8.1	550	11.0	<2	1	2.1
	07-14-2009	1000	2.4	0.5	10.6	7.5	593	8.0	<2	1	0.8
	08-12-2009	1135	2.6	0.5	11.7	8.0	626	6.3	<2	–	–
	09-09-2009	0920	4.9	1.0	11.5	7.7	678	4.6	<2	1	0.8
	06-01-2010	1630	4.2	0.8	9.5	8.5	584	4.1	<2	4	1.7
	07-13-2010	1640	7.9	1.6	11.0	8.1	594	8.0	<2	1	1.3
	09-07-2010	1555	13.9	2.8	11.5	8.0	651	7.4	<2	2	0.7
	06-07-2011	1420	9.1	1.8	12.1	8.1	531	7.2	<2	1	1.5
	07-20-2011	1645	4.9	1.0	9.6	8.0	571	10.2	<2	<1	1.0
	09-05-2011	1600	13.6	2.7	10.5	8.2	580	5.5	<2	<1	1.0

Table 3.   Physical field parameters, suspended sediment, and dissolved organic carbon measured at surface water sites in Kantishna Hills, Alaska, 2008–2011.—Continued

[Number below constituent in parentheses is used by U.S. Environmental Protection Agency and U.S. Geological Survey to identify parameters in computer databases. **Abbreviations:** ft^3/s, cubic foot per second; (ft^3/s)/mi^2, cubic foot per second per square mile; mg/L, milligram per liter; μS/cm at 25°C, microsiemens per centimeter at 25 degrees Celsius; °C, degrees Celsius; FNU, Formazin Nephelometric Unit; mi^2, square mile; <, less than; –, not measured; E, estimated]

Site No.	Date	Time	Streamflow (ft^3/s) (00061)	Unit streamflow [(ft^3/s)/mi^2]	Dissolved oxygen (mg/L) (00300)	pH (units) (00400)	Specific conductance (μS/cm at 25°C) (00095)	Water temperature (°C) (00010)	Turbidity (FNU) (63680)	Suspended sediment (mg/L) (80154)	Dissolved organic carbon (mg/L) (00681)
					Moose Creek (drainage area 92.7 mi^2)						
10	06-11-2008	1425	233	2.5	11.7	8.0	164	5.7	2.5	7	3.7
	07-15-2008	1640	145	1.6	–	7.8	182	11.5	–	2	1.5
	09-09-2008	1145	187	2.0	12.2	8.1	210	4.7	<2	1	1.1
					Eureka Creek (drainage area 5.0 mi^2)						
11	06-13-2008	1245	14.0	2.8	11.5	8.1	389	5.6	<2	1	5.1
	07-14-2008	1150	7.7	1.5	–	8.0	538	6.0	<2	1	2.3
	08-22-2008	1750	17.0	3.4	12.6	8.1	478	5.9	<2	–	–
	09-09-2008	1510	9.4	1.9	11.8	8.3	568	6.1	<2	1	1.3
	06-03-2009	1655	4.3	0.9	9.7	8.3	467	10.0	6.5	1	4.0
	07-13-2009	1900	2.8	0.6	10.0	8.1	671	13.8	<2	2	1.7
	08-10-2009	1840	4.1	0.8	10.9	8.1	690	9.3	<2	–	–
	09-08-2009	1500	4.4	0.9	10.4	8.2	696	7.3	<2	1	1.4
	05-31-2010	1930	5.3	1.1	9.8	8.1	480	4.7	<2	1	3.0
	07-12-2010	1930	6.5	1.3	9.5	8.2	523	9.9	<2	<1	2.7
	09-06-2010	1805	8.2	1.6	11.0	8.2	573	7.2	<2	<1	2.0
	06-06-2011	1910	8.1	1.6	11.6	8.1	454	7.8	3.7	8	3.7
	07-20-2011	1900	6.7	1.3	10.3	8.2	549	9.0	<2	1	2.0
	09-05-2011	1755	10.3	2.1	11.1	8.3	544	6.0	<2	<1	1.7
					Slate Creek (drainage area 0.5 mi^2)						
12	06-12-2008	1600	–	–	–	6.9	210	–	–	24	3.0
	07-13-2008	1231	1.1	2.2	11.4	7.2	286	4.8	<2	11	2.2
	08-21-2008	1150	1.5	3.0	11.4	6.7	322	4.8	<2	–	–
	09-11-2008	0935	0.6	1.2	11.1	6.7	339	5.5	<2	6	1.6
	06-02-2009	1850	2.0	4.0	10.6	7.2	255	4.8	14	12	3.1
	07-28-2009	1750	0.4	0.8	10.0	6.6	355	10.0	2.1	5	1.5
	08-11-2009	1750	0.3	0.6	11.2	6.9	361	8.5	5.3	–	–
	09-09-2009	1050	0.6	1.2	10.9	6.6	286	6.1	<2	7	1.7
	06-01-2010	1530	0.5	1.0	9.2	7.8	289	4.0	4.4	15	6.7
	07-13-2010	1515	2.1	4.2	11.2	6.2	240	6.2	4.1	6	2.7
	09-07-2010	1509	1.5	3.0	11.2	7.1	337	6.7	<2	16	1.6
	06-07-2011	1520	1.0	2.0	12.5	6.5	259	4.5	<2	6	2.6
	07-21-2011	1405	0.9	1.8	8.0	6.3	344	8.0	<2	7	1.8
	09-06-2011	1345	0.8	1.6	10.8	6.6	367	6.0	<2		1.9

Table 3.   Physical field parameters, suspended sediment, and dissolved organic carbon measured at surface water sites in Kantishna Hills, Alaska, 2008–2011.—Continued

[Number below constituent in parentheses is used by U.S. Environmental Protection Agency and U.S. Geological Survey to identify parameters in computer databases. **Abbreviations:** ft^3/s, cubic foot per second; $(ft^3/s)/mi^2$, cubic foot per second per square mile; mg/L, milligram per liter; µS/cm at 25°C, microsiemens per centimeter at 25 degrees Celsius; °C, degrees Celsius; FNU, Formazin Nephelometric Unit; mi^2, square mile; <, less than; –, not measured; E, estimated]

Site No.	Date	Time	Streamflow (ft^3/s) (00061)	Unit streamflow $[(ft^3/s)/mi^2]$	Dissolved oxygen (mg/L) (00300)	pH (units) (00400)	Specific conductance (µS/cm at 25°C) (00095)	Water temperature (°C) (00010)	Turbidity (FNU) (63680)	Suspended sediment (mg/L) (80154)	Dissolved organic carbon (mg/L) (00681)
— Eldorado Creek (drainage area 12.5 mi²)										
13	06-13-2008	1030	9	0.7	11.7	8.1	443	4.5	<2	2	5.0
	07-14-2008	1530	11	0.9	–	8.0	505	6.0	<2	1	2.5
	09-09-2008	1400	19	1.5	12.1	8.2	488	4.6	<2	1	1.3
	06-01-2010	0900	7.2	0.6	10.4	8.2	528	2.7	<2	1	4.4
	07-12-2010	2120	12.2	1.0	9.6	8.0	499	9.1	6.4	5	2.6
	09-06-2010	1700	19.1	1.5	11.0	8.0	467	8.9	<2	1	1.4
Friday Creek (drainage area 1.5 mi²)										
14	06-13-2008	1430	1.3	0.9	11.6	8.2	447	5.2	<2	1	5.6
	07-14-2008	1325	1.3	0.9	–	8.1	550	5.5	<2	1	3.0
	08-24-2008	1620	3.1	2.1	12.9	8.1	482	5.5	<2	–	–
	09-09-2008	1740	1.7	1.1	12.0	8.4	588	5.0	3.7	1	1.7
	06-03-2009	1900	0.3	0.2	10.5	8.3	490	8.4	<2	1	4.6
	07-13-2009	1810	0.1	0.1	10.8	8.3	657	11.1	<2	1	1.9
	08-11-2009	2100	0.2	0.1	13.4	8.2	717	3.8	<2	–	–
	09-08-2009	1610	0.4	0.3	10.8	8.3	688	6.3	<2	1	1.6
Moose Creek (drainage area 152 mi²)										
15	06-11-2008	1750	289	1.9	11.3	8.0	221	7.4	<2	6	4.0
	07-14-2008	0908	202	1.3	12.4	7.7	251	6.3	<2	1	1.7
	09-08-2008	1755	234	1.5	11.2	7.5	279	7.7	<2	2	1.2
	06-02-2009	0905	180	1.2	11.3	7.7	237	5.6	<2	7	2.3
	07-14-2009	1715	105	0.7	10.0	7.7	278	14.3	<2	1	1.0
	09-08-2009	1750	169	1.1	10.2	7.8	262	8.5	<2	2	1.0
	05-31-2010	1730	92	0.6	8.5	7.7	290	11	<2	2	1.9
	07-12-2010	1720	181	1.2	8.9	7.9	256	13.2	<2	<1	1.6
	09-06-2010	1545	313	2.1	10.8	8.0	273	8.9	<2	2	1.2
	06-06-2011	1730	174	1.1	10.8	7.8	267	11.2	<2	2	1.7
	07-20-2011	2045	235	1.5	9.7	7.9	251	11.6	<2	6	2.0
	09-06-2011	0735	320	2.1	12.0	8.2	296	4.3	<2	2	1.2

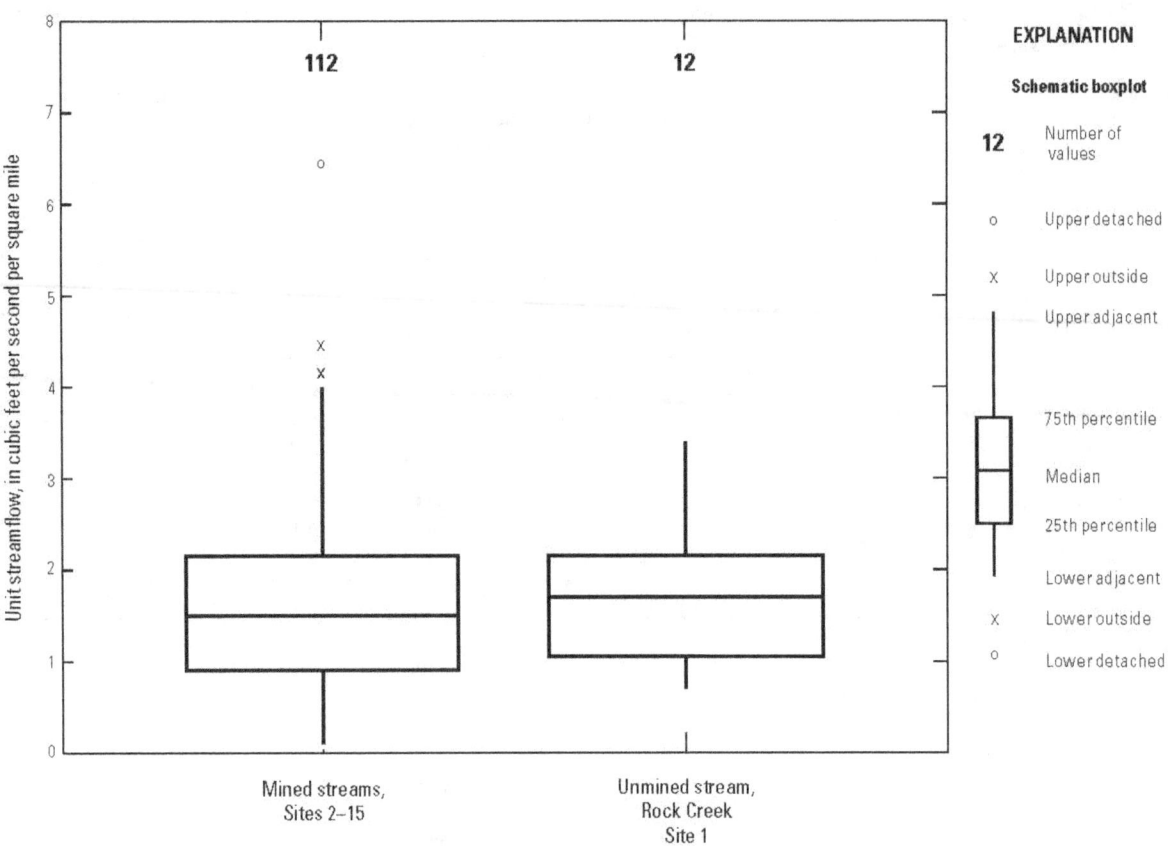

Figure 18.   Unit streamflow measured at all sites and Rock Creek, Kantishna Hills, Alaska, June 10, 2008–September 6, 2011.

Suspended Sediment and Turbidity

Sediment in rivers is transported in suspension (water column) and along the streambed (bedload). Suspended sediment generally consists of fine particles such as clay, silt, and fine sand that are held in suspension by the turbulence of flowing water. Bedload consists of coarse sediment particles such as sands, gravels, and sometimes boulders that are transported along or near the streambed. Turbidity, a surrogate for suspended sediment in water, is caused by suspended and dissolved matter such as clay, silt, finely divided organic matter, plankton, and other microscopic organisms. Suspended-sediment concentrations and turbidity values are affected by a number of factors: the amount of precipitation and runoff, intensity and duration of storms, and

geomorphic structure of the channel. High suspended-sediment concentrations and turbidity values reduce light penetration and photosynthesis, smother benthic habitats, and interfere with fish feeding activities.

Examination of the historical suspended sediment and turbidity data indicated no differences at Rock Creek through the 1980s and 1990s (appendixes A, D, E, and F). High turbidity values and suspended-sediment concentrations were noted in 1982, 1983, and 1984 (appendixes A, B, C, D, and E) where active mining was occurring: Caribou Creek (sites 2, 4, and 5), East Fork Glen Creek (site 7), Glen Creek (site 9), Eureka Creek (site 11), Slate Creek (site 12), Friday Creek (site 14), and Moose Creek (near site 15). However, at these same sites, turbidity values were relatively low in the 1990s. These lower turbidity values probably are due to several factors; all mining had stopped in 1985 and thus there was no active earth moving, reclamation projects had started in several watersheds, and natural revegetation had begun.

During the 2008–11 study, the water samples collected for suspended sediment and field parameter readings for turbidity were collected during average or low streamflows. No samples were collected during high streamflows when suspended-sediment concentrations and turbidity values would be expected to be higher. Based on the data collected, suspended-sediment concentrations and turbidity values were low and sometimes below detection limits (table 3). As noted, several sites were documented with high suspended-sediment concentrations and turbidity values during average or low streamflows in the 1980s. Thus, the differences between the 1980s and this study indicate changes have occurred in these watersheds.

To determine turbidity (and indirectly suspended sediment) characteristics at high flows, instruments were installed at six sites to collect continuous turbidity data in 2009: Rock Creek (site 1), Caribou Creek (sites 2 and 6), and three sites on Glen Creek (miscellaneous sites 1 and 2, site 9). Three Glen Creek sites were selected because reclamation was taking place in 2009, which offered the opportunity to assess sediment characteristics downstream of the reclamation. All instruments performed well except the turbidity probe at Glen Creek near the mouth (site 9), which was damaged during a high flow event.

Turbidity data collected from Rock Creek, the control site, and Caribou Creek (site 6), the farthest downstream site mined on Caribou Creek, showed nearly identical patterns in turbidity throughout the summer of 2009 (fig. 19). The periods of high turbidity were somewhat coincident with rainfall at Wonder Lake that increased stage and the amount of water in the stream. The similar patterns of these two streams indicate that changes in the Caribou Creek watershed are reducing turbidity. These changes include no mining or active disturbance of the land, reclamation projects that have reduced sediment sources (flattening tailings piles), and the natural revegetation of disturbed areas that has taken place since 1985.

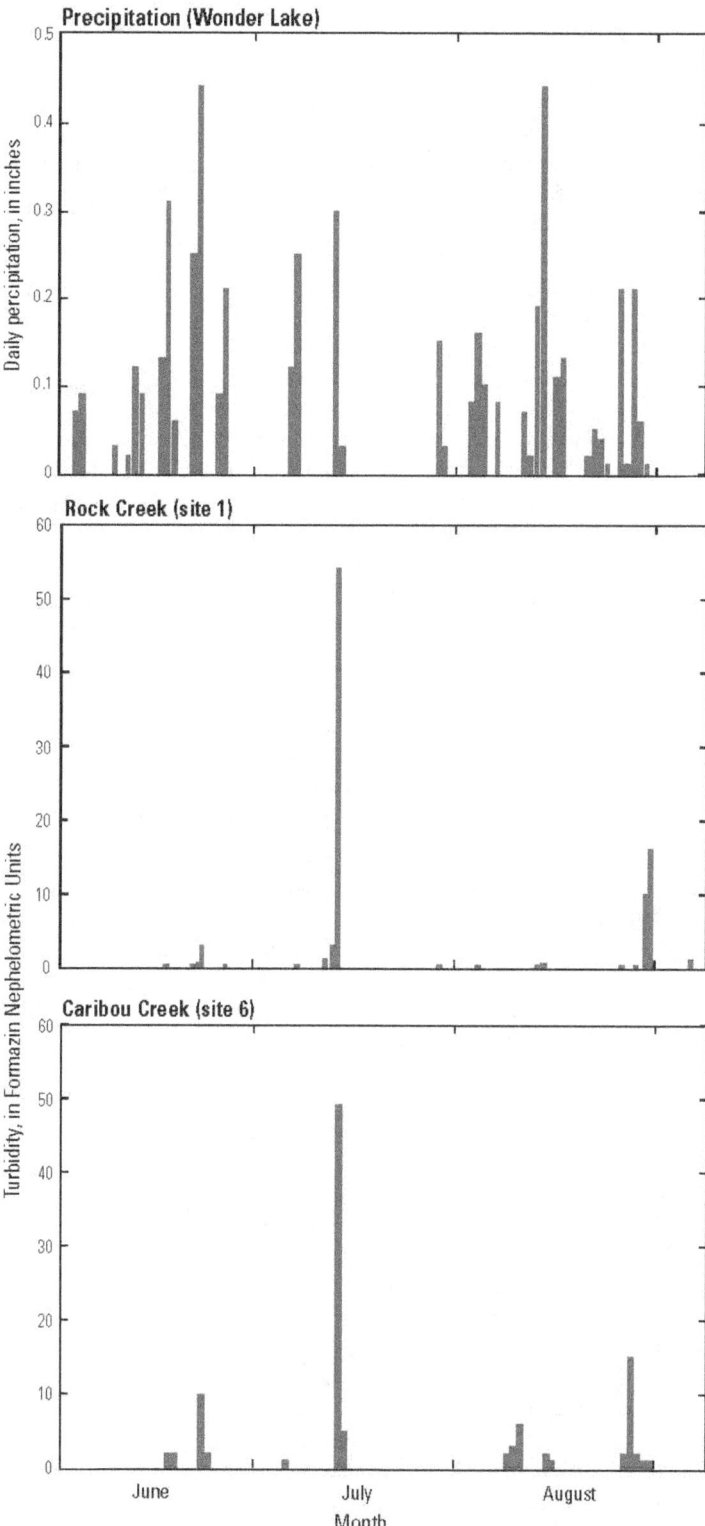

Figure 19.   Average daily turbidity at Rock Creek and Caribou Creek site 6 and daily precipitation at Wonder Lake, Kantishna Hills area, Alaska, June 3–September 9, 2009.

In the Glen Creek watershed, miscellaneous site 1 was about 0.25 mi downstream of the reclamation project and miscellaneous site 2 was about halfway between the miscellaneous site 1 and site 9 near the mouth of Glen Creek (fig. 5). At miscellaneous site 1, the highest turbidity value recorded was 55 formazin Nephelometric units (FNU) (fig. 20) and is not considered an indicator of high suspended sediment. Generally, values of turbidity at miscellaneous site 2 were always lower than miscellaneous site 1. Glen Creek is a gaining stream and the increase in flow between miscellaneous sites 1 and 2 is from groundwater that likely dilutes the turbidity from miscellaneous site 1. In August, turbidity readings less than 15 FNUs were recorded at miscellaneous site 2 and may have been caused by rainfall. However, no readings of turbidity greater than the detection limit were recorded at miscellaneous site 1, which may have been a result of a malfunctioning probe.

Water Temperature

Water temperature determines the amount of oxygen water can contain when at equilibrium with the atmosphere and controls the metabolic and growth rates of fish. In 2008, the five sites along Caribou Creek (sites 2–6) were sampled, and water temperature showed an increase from the upstream site (site 2) to the downstream site (site 6). At Caribou Creek sites 2 and 6 the highest water temperatures were measured in July 2009, corresponding with the warm summer air temperatures. Water temperatures were as high as 14.2°C during this July period (table 3). The lowest water temperatures were measured at the beginning of the runoff period in early June, reflecting snowmelt, and were as low as 2.7°C (table 3).

Probes recorded daily water temperatures at Rock Creek, Caribou Creek sites 2 and 6, and at three sites on Glen Creek in 2009 (fig. 5). At all sites, water temperature generally corresponded with air temperature (fig. 21) throughout the summer. Water temperatures were as high as 15°C at the downstream Caribou Creek site (site 6) and as noted, were warmer than the upstream Caribou Creek site (site 2). In Glen Creek, water temperatures were relatively constant throughout the stream reach, ranging from approximately 3 to 9°C.

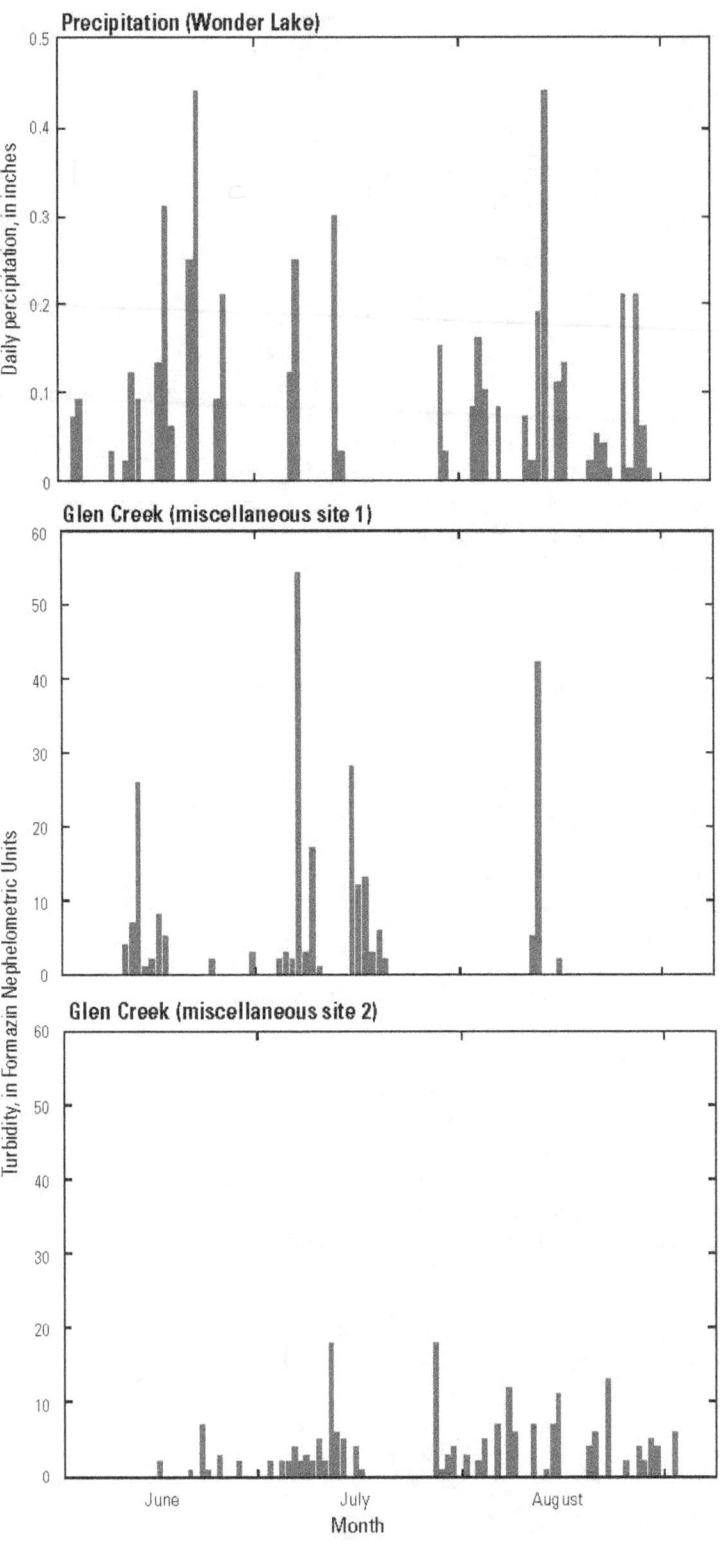

Figure 20. Average daily turbidity at Glen Creek miscellaneous sites 1 and 2, Kantishna Hills area, Alaska, June 3–September 9, 2009.

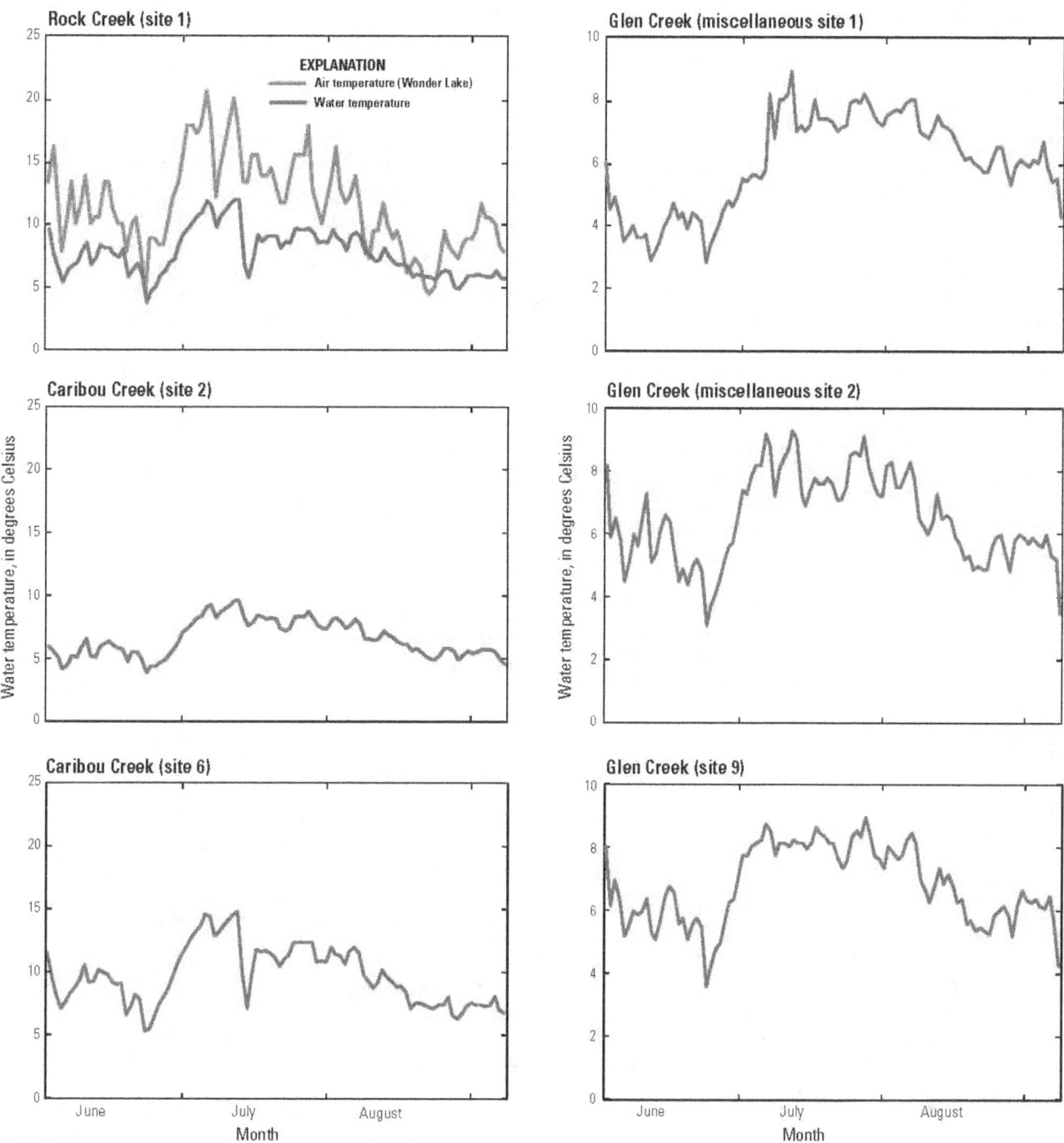

Figure 21. Average daily air temperature at Wonder Lake and average daily water temperatures at Rock Creek site 1, Caribou Creek sites 2 and 6, Glen Creek miscellaneous sites 1 and 2, and Glen Creek site 9, Kantinsha Hills area, Alaska, June 3–September 3, 2009.

Specific Conductance

Specific conductance is a measure of the ability of water to conduct an electric current and can be used to indicate the concentrations of dissolved solids, or ions, in the water. As the concentration of ions in solution increases or decreases, so does the conductance of the solution. Commonly, a statistical relation can be developed between specific conductance and the ionic components making up the dissolved solids in water. During low flow, the conductance of stream water generally is the highest, indicating a greater component of groundwater in the total flow. Groundwater has greater potential to dissolve minerals than does rainwater or snowmelt, having spent more time in contact with rocks and soil materials. Periods of relatively low specific conductance in stream water reflect runoff of rain or snowmelt, which typically contain small amounts of dissolved ions.

Within a given watershed, there is a constantly changing relation between streamflow and solute concentration during storm runoff. This relation may be represented in a circular pattern that is referred to as hysteresis. The pattern of the relation between discharge and concentration is circular because the total concentration levels of solutes vary during different periods of a storm. This variation in concentration is not merely the product of dilution, but is also caused by end-member mixing; that is, the mixing of the different components of discharge from multiple sources such as groundwater and surface water. In addition, because mining operations excavate and expose large amounts of overburden, specific conductance will usually be higher because of the mineralized drainage from tailing piles and disturbed lands into the receiving stream. Thus, a strong correlation between specific conductance and streamflow was not expected in the Kantishna Hills study area.

During the 2008–11 study, specific conductance was highest at streams draining mined lands (Caribou Creek, Glen Creek, Eldorado Creek, Eureka Creek, and Friday Creek) reflecting the disturbed nature of these watersheds (fig. 5; table 3). The lowest values of specific conductance were measured at Moose Creek. The upstream site on Moose Creek (site 10) reflects input from Glen Creek, whereas the downstream site (site 15) reflects the input from Eldorado Creek, Eureka Creek, and Friday Creek, and thus had a slightly higher specific conductance.

Specific conductance data were collected at some sites in the 1980s and 1990s (appendixes A, E, F). The historical datasets from the 1980s are limited to only a few samples at several sites; therefore, it was determined that no valid comparisons could be made to the 1990s or current study data. However, some comparisons could be made between the 1990s data and the data collected during this study. Comparison of the averages of the 1990s data (appendix F) with the averages of the specific conductance data collected during this study indicated that specific conductance has increased at Rock Creek (site 1), Caribou Creek (site 2),

Caribou Creek (site 6), Friday Creek (site 14), and Moose Creek (site 15). At Eureka Creek (site 11), specific conductance has remained fairly constant.

Specific conductance data were collected on a continuous basis in 2009 at the same six sites continuously monitored for turbidity and water temperature. Generally, specific conductance decreased during rainfall that resulted in higher streamflow (fig. 22). As streamflow decreased, specific conductance gradually increased. Values of specific conductance were as low as 125 µS/cm in Rock Creek and as high as 780 µS/cm at Caribou Creek site 2. The specific conductance at Caribou Creek site 6 was lower than Caribou Creek site 2, and showed more noticeable changes during rainfall events, likely due to the inflow of additional water with a lower specific conductance. In the Glen Creek watershed, specific conductance remained about the same from miscellaneous site 2 to the mouth of Glen Creek (site 9). In the upper part of the watershed, miscellaneous site 1 exhibited different characteristics that may have been caused by the reclamation work upstream of the site in 2009.

pH

The pH of water is a measure of the hydrogen-ion activity and typically ranges from 0 (acidic) to 14 (alkaline) standard units. The pH of natural river water typically ranges between 6.5 and 8.0 standard units (Hem, 1985). During the study period, with the exception of Slate Creek, measured values of pH for all sites ranged from 7.5 to 8.5 (table 3), which is within an acceptable range for cold-water fish growth and survival. Measurements of pH at Slate Creek were collected downstream of the mineralized zone and ranged from 6.2 to 7.8. Low pH values (2.7–5.8) have been measured in water that directly drains the antimony deposit (Eppinger and others, 2000). Historical measurements of pH (appendixes A, B, C, E, F) indicated no changes from the present levels.

Dissolved Oxygen

The dissolved-oxygen concentration in a stream is controlled by several factors, including water temperature, air temperature and atmospheric pressure, photosynthetic or respiratory activity of stream biota, and the quantity of organic matter present (Hem, 1985). Salmon and other species of fish require well-oxygenated water at every life stage, as do many forms of aquatic invertebrates. Young fish tend to be more susceptible to oxygen deficiencies than adults. All streams during the study period were well oxygenated and concentrations ranged from 8.0 to 13.8 mg/L (table 3), the range of values being nearly the same at all sites. At times during the month of July, when air temperatures were highest, there was a corresponding decrease in dissolved oxygen. Historical concentrations of dissolved oxygen (appendix F) were similar in range to the results of the present study.

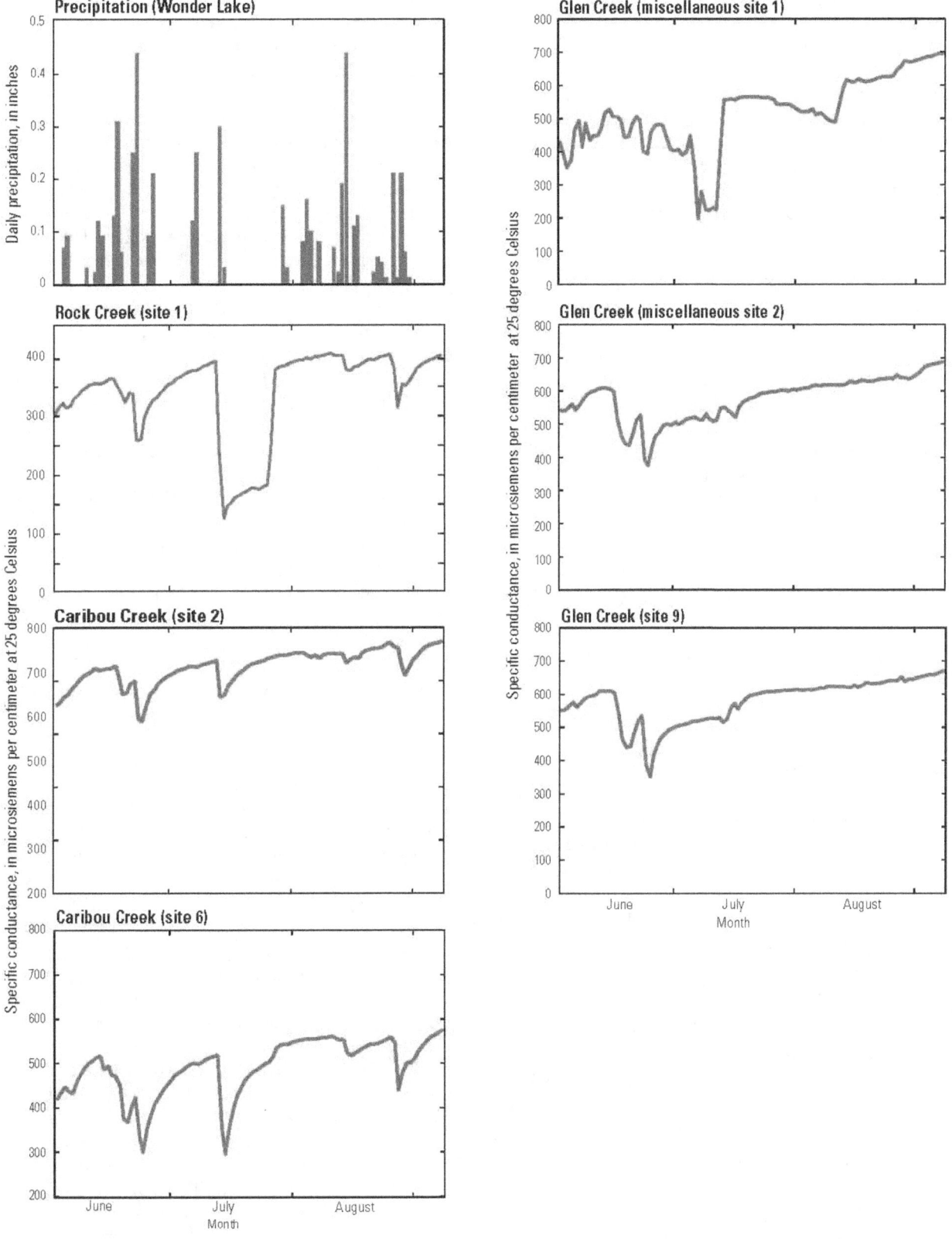

Figure 22. Average daily rainfall at Wonder Lake and average daily specific conductance at Rock Creek, Caribou Creek site 2, Caribou Creek site 6, Glen Creek miscellaneous sites 1 and 2, and Glen Creek site 9, Kantishna Hills area, Alaska.

Organic Carbon

Dissolved organic carbon (DOC) is commonly a major component of organic matter in aquatic ecosystems. DOC is defined as organic carbon in the filtrate (dissolved and colloidal phases) that passes through a 0.45-μm pore-size filter (Aiken and Cotsaris, 1995). Generally, DOC is in greater abundance than particulate organic carbon and accounts for about 90 percent of the total organic carbon of most waters (Aiken and Cotsaris, 1995). During the winter, organic carbon accumulates in the soil and then is removed during snowmelt. At all sites, except Slate Creek, DOC concentrations were highest in samples collected in June 2008. Based on the measured discharges, the June 2008 sampling probably took place at or near the time of peak snowmelt, whereas the June sampling during 2009–11 probably took place after peak snowmelt. The highest concentration of DOC during 2008–11 was 6.7 mg/L, at Slate Creek (table 3). During the remainder of summer and autumn, concentrations of DOC were relatively low, ranging from 0.5 to 3.5 mg/L. These relatively low concentrations of DOC are an indicator of the lack of well developed soils in the Kantishna Hills area.

Alkalinity

Alkalinity is a measure of the capacity of the substances dissolved in water to neutralize acid. In most natural waters, alkalinity is produced mainly by bicarbonate and carbonate ions, which are formed when carbon dioxide or carbonate rocks dissolve in water (Hem, 1985). The lowest alkalinity concentrations (49–85 mg/L) (table 4) were measured at the two Moose Creek sites (sites 10 and 15) and at Slate Creek (site 12) (64–99 mg/L). Relatively high alkalinity concentrations were measured at Caribou Creek (site 2), East Fork Glen Creek, Eureka Creek, Eldorado Creek, and Friday Creek sites (87–194 mg/L). The range of pH measured at all sites indicates that all the alkalinity can be attributed to dissolved bicarbonate (Hem, 1985). Alkalinity concentrations at all sites are considered average to high and indicate a high buffering capacity.

Alkalinity data were collected in the 1980s and 1990s (appendixes A, B, C, E, F) at several sites. Although only a few samples were collected in the 1980s, comparison of data from the 1990s to 2008–11 indicated that alkalinity has increased slightly at Rock Creek and Caribou Creek (site 6), but has remained unchanged at Glen, Eureka, Friday, and Moose Creeks.

Dissolved Solids and Major Ions

Major dissolved constituents in streams consist of inorganic minerals derived primarily from soil and rock weathering. Dissolved cations that constitute a majority of the dissolved solid content in natural waters are calcium, magnesium, sodium, and potassium; the major anions are usually represented by sulfate, chloride, fluoride, nitrate, and those making up the alkalinity (carbonate and bicarbonate) (Hem, 1985). Streams draining watersheds that consist of rocks and soils having insoluble minerals contain lower concentrations of dissolved solids than streams draining watersheds containing easily dissolved minerals. As previously noted, mining operations increase concentrations of dissolved solids and ions by discharging highly mineralized drainage from tailing piles and disturbed lands into the receiving stream. Analyses of the water samples collected in this study indicated that total dissolved solid concentrations are highest (average of 472 mg/L) at Caribou Creek (site 2), but decrease to an average of 317 mg/L at Caribou Creek (site 6) (table 4). Average concentrations of dissolved solids also were similar at the other mined sites: Glen Creek (site 9) (414 mg/L), Eureka Creek (site 11) (367 mg/L), Eldorado Creek (site 13) (305 mg/L), and Friday Creek (site 14) (391 mg/L). The lowest average concentrations were at the Moose Creek sites (sites 10 and 15) (115 and 164 mg/L, respectively) and at Rock Creek (site 1) (221 mg/L).

Comparison of dissolved solids data collected in the 1990s (appendix F) with dissolved solids data collected during this study indicated increasing concentrations at a number of sites: Rock Creek, Caribou Creek (site 6), Glen Creek (site 9), Eureka Creek, Friday Creek, and Moose Creek (site 15). The increases ranged from about 20 percent at Moose Creek (average of 137–164 mg/L) to about 53 percent at Friday Creek (average of 256–391 mg/L). Because the dissolved solids concentrations have increased at Eureka Creek and Friday Creek, a corresponding increase in concentration at Moose Creek (site 15) would be expected.

The cations calcium and magnesium are common alkaline-earth metals that are essential elements in plant and animal nutrition. Both elements are major anion components in most natural waters (Hem, 1985). In the samples collected in this study, concentrations of calcium ranged from 21.4 to 129 mg/L and those of magnesium ranged from 5.0 to 40.7 mg/L (table 4). The highest concentrations of the constituents were at Glen Creek (site 9) (calcium) and Eureka Creek (site 11) (magnesium). Sodium and potassium are present in most natural waters, but usually in low concentrations in rivers. Sodium concentrations ranged from 0.45 to 5.44 mg/L and potassium concentrations ranged from 0.22 to 2.13 mg/L among all sites (table 4).

For anions, bicarbonate (HCO_3) concentrations ranged from 60 to 252 mg/L, with the highest concentrations detected at Eureka Creek and Friday Creek. Sulfate and silica, which are dissolved from rocks and soils, are the next most abundant anions, with concentrations from 27.4 to 330 mg/L for sulfate, and 3.08 to 11.3 mg/L for silica. The highest concentrations of sulfate were detected at Caribou Creek (site 2), East Fork Glen Creek (site 7), and Glen Creek (site 9). Chloride and fluoride concentrations were less than or equal to 14.6 and 0.13 mg/L, respectively, at all sites (table 4).

Table 4. Concentrations of major ions and dissolved solids in water samples collected in Kantishna Hills area, Alaska, 2008–11.

[Number below constituent in parentheses is used by U.S. Environmental Protection Agency and U.S. Geological Survey to identify parameters in computer databases. All values in milligrams per liter. **Site No.:** Site locations are shown in figure 2. **Bold** value indicates average. **Abbreviations:** E, estimated value; <, less than reporting level; –, no data or constitutent not measured]

Site No.	Date	Time	Alkalinity (39086)	Bicar-bonate (00453)	Calcium (00915)	Chloride (00940)	Total dissolved solids (70300)	Fluoride (00950)	Mag-nesium (00925)	Potas-sium (00935)	Silica (00955)	Sodium (00930)	Sulfate (00945)
						Rock Creek							
1	06-10-2008	1200	70	86	36.1	E0.11	177	<0.12	10.3	1.24	4.16	0.91	58.0
	07-13-2008	1410	88	107	44.2	E0.11	211	E0.06	12.5	1.23	5.27	1.23	76.1
	09-10-2008	0955	92	113	48.6	0.13	234	<0.12	13.7	1.29	5.24	1.32	86.8
	06-03-2009	1120	81	105	43.4	0.21	200	E0.06	11.8	1.40	5.01	1.23	69.6
	07-28-2009	1620	106	138	53.8	0.13	254	E0.06	15.9	1.54	5.90	1.50	93.0
	09-10-2009	1000	112	146	57.6	0.14	269	E0.07	16.3	1.38	5.54	1.40	101
	06-01-2010	1320	90	117	51.7	0.21	242	0.10	14.1	1.44	5.15	1.58	86.4
	07-13-2010	1355	83	108	46.1	E0.10	197	0.10	13.1	1.18	5.43	1.31	71.7
	09-07-2010	1310	90	117	51.4	E0.12	232	E0.04	14.5	1.25	5.48	1.23	85.6
	06-07-2011	1340	74	97	39.8	0.12	182	<0.04	11.5	1.27	4.80	1.14	63.9
	07-21-2011	1130	99	129	50.5	0.11	218	0.05	14.4	1.28	5.76	1.41	81.6
	09-06-2011	1245	91	118	46.0	0.15	233	0.05	13.7	1.16	5.51	1.14	81.6
			90	**115**	**47.4**	**–**	**221**	**–**	**13.5**	**–**	**–**	**–**	**79.6**
						Caribou Creek							
2	06-12-2008	1120	92	112	75	E0.09	376	<0.12	23.5	1.40	3.08	1.07	188
	07-15-2008	1019	130	158	96.9	E0.09	498	<0.12	29.9	1.56	3.75	1.53	244
	09-10-2008	1645	123	150	89.3	E0.11	453	<0.12	28.0	1.65	–	1.65	219
	06-03-2009	1045	113	147	93.4	0.20	447	E0.05	28.3	1.63	3.72	1.50	227
	07-28-2009	1305	141	183	110	0.12	534	<0.08	33.4	1.82	4.51	1.79	271
	09-09-2009	1230	140	182	113	0.15	564	E0.06	35.1	1.74	4.32	1.59	289
	06-01-2010	1200	115	150	105	0.15	479	E0.06	30.4	1.74	4.00	1.95	242
	07-13-2010	1120	121	157	97.7	E0.09	443	0.09	30.6	1.58	3.99	1.69	227
	09-07-2010	1223	122	159	97.3	E0.10	458	<0.08	30.0	1.62	4.25	1.54	230
			122	**155**	**97.5**	**–**	**472**	**–**	**29.9**	**–**	**–**	**–**	**237**
3	06-12-2008	0940	84	102	56.3	E0.09	306	<0.12	21.0	1.25	3.38	0.89	139
	07-15-2008	1120	117	142	73.6	E0.10	396	<0.12	27.0	1.43	4.09	1.35	185
	09-10-2008	1530	119	145	76.1	E0.10	419	<0.12	29.7	1.39	4.24	1.48	171
			107	**130**	**68.7**	**–**	**374**	**–**	**25.9**	**–**	**–**	**–**	**165**
4	06-10-2008	1645	79	96	51.4	E0.10	297	<0.12	19.1	1.29	3.57	0.9	125
	07-15-2008	1240	106	129	68.6	E0.10	349	<0.12	24.0	1.38	4.6	1.37	161
	09-10-2008	1410	112	137	72.9	E0.11	378	<0.12	27.4	1.45	4.63	1.58	180
			99	**121**	**64.3**	**–**	**341**	**–**	**23.5**	**–**	**–**	**–**	**155**
5	06-10-2008	1345	73	89	47.0	E0.10	272	<0.12	17.7	1.25	3.8	0.91	115
	07-13-2008	1641	95	116	57.5	E0.09	358	<0.12	21.1	1.30	5.0	1.34	141
	09-10-2008	1300	106	130	67.8	E0.11	365	<0.12	25.9	1.41	4.9	1.62	173
			91	**112**	**57.4**	**–**	**332**	**–**	**21.6**	**–**	**–**	**–**	**143**
6	06-10-2008	1520	76	92	46.0	E0.09	251	<0.12	17.2	1.29	3.91	1.01	112
	07-13-2008	1531	90	110	56.0	E0.09	284	<0.12	19.7	1.26	5.20	1.43	130
	09-10-2008	1150	104	127	65.6	E0.11	349	<0.12	25.1	1.39	5.05	1.72	166
	06-03-2009	1230	82	107	54.1	0.19	292	E0.08	19.9	1.40	4.63	1.33	128
	07-28-2009	1450	114	148	69.6	E0.11	379	<0.08	27.1	1.66	5.85	1.91	175
	09-09-2009	1730	117	152	75.4	0.13	402	E0.06	27.8	1.48	5.47	1.73	194
	06-01-2010	1420	88	114	63.8	0.18	309	E0.07	23.1	1.44	4.56	1.58	154
	07-13-2010	1240	85	110	58.0	E0.08	283	E0.08	21.4	1.29	5.45	1.53	128
	09-07-2010	1400	100	130	66.1	E0.11	333	E0.06	24.0	1.29	5.45	1.47	160
	06-07-2011	1240	79	103	52.1	–	260	<0.04	19.6	1.39	4.53	1.41	–
	07-21-2011	1230	100	130	61.4	0.09	306	0.04	22.7	1.30	5.82	1.76	143
	09-06-2011	1140	100	130	59.1	0.14	361	0.04	22.9	1.26	5.56	1.47	145
			95	**121**	**60.6**	**–**	**317**	**–**	**22.5**	**–**	**–**	**–**	**150**

Table 4.   Concentrations of major ions and dissolved solids in water samples collected in Kantishna Hills area, Alaska, 2008–11. — Continued

[Number below constituent in parentheses is used by U.S. Environmental Protection Agency and U.S. Geological Survey to identify parameters in computer databases. All values in milligrams per liter. **Site No.:** Site locations are shown in figure 2. **Bold** value indicates average. **Abbreviations:** E, estimated value; <, less than reporting level; –, no data or constitutent not measured]

Site No.	Date	Time	Alkalinity (39086)	Bicarbonate (00453)	Calcium (00915)	Chloride (00940)	Total dissolved solids (70300)	Fluoride (00950)	Magnesium (00925)	Potassium (00935)	Silica (00955)	Sodium (00930)	Sulfate (00945)
						East Fork Glen Creek							
7	6-12-2008	1445	121	148	80.7	E0.09	375	<0.12	19.5	1.11	3.82	0.71	157
	7-15-2008	1505	147	180	101	E0.07	460	<0.12	24.8	1.05	4.38	0.92	219
	9-11-2008	0930	150	184	129	0.14	667	<0.12	40.3	1.29	4.76	1.19	330
			139	**171**	**103.6**	**–**	**501**	**–**	**28.2**	**–**	**–**	**–**	**235**
						West Fork Glen Creek							
8	06-12-2008	1300	65	79	46.9	<0.12	211	<0.12	9.0	0.94	3.52	0.45	85.4
	07-15-2008	1405	119	145	85.1	E0.07	351	<0.12	15.3	0.95	4.88	0.75	163
	09-11-2008	1535	106	129	85.7	0.13	388	<0.12	17.5	1.04	5.02	1.07	185
			97	**118**	**72.6**	**–**	**317**	**–**	**13.9**	**–**	**–**	**–**	**144**
						Glen Creek							
9	06-11-2008	1140	85	103	61.2	E0.09	316	<0.12	16.5	0.96	4.05	0.88	130
	07-15-2008	1050	115	141	83.8	E0.10	398	<0.12	22.9	1.05	4.89	1.24	184
	09-09-2008	1020	116	142	90.3	0.12	448	<0.12	25.9	1.03	4.87	1.32	218
	06-02-2009	1635	107	139	84.2	0.20	388	E0.04	22.5	1.12	4.81	1.20	186
	07-14-2009	1000	113	147	83.1	0.13	412	<0.08	22.9	0.95	4.89	1.36	199
	09-09-2009	920	134	174	102	0.13	496	<0.08	27.7	1.04	5.36	1.36	239
	06-01-2010	1630	104	135	90.1	0.15	413	E0.06	24.1	0.95	4.14	1.14	207
	07-13-2010	1640	115	150	89.8	E0.09	420	0.08	25.7	1.00	5.12	1.25	195
	09-07-2010	1555	124	161	97.9	E0.11	460	<0.08	26.8	1.02	5.45	1.27	229
	06-07-2011	1420	102	133	80.7	0.12	381	<0.04	22.4	1.10	4.71	1.21	178
	07-20-2011	1645	120	156	85.8	0.12	412	<0.04	24.0	0.98	5.45	1.34	192
	09-05-2011	1600	122	159	89.1	0.15	420	<0.04	25.0	0.99	5.44	1.20	210
			113	**145**	**86.5**	**–**	**414**	**–**	**23.9**	**–**	**–**	**–**	**197**
						Moose Creek							
10	06-11-2008	1425	49	60	21.4	0.60	108	E0.06	5.0	0.56	7.74	2.43	27.8
	07-15-2008	1640	59	72	24.9	0.77	111	E0.09	5.3	0.53	10.00	3.21	27.4
	09-09-2008	1145	64	78	27.2	0.64	127	E0.06	6.1	0.52	9.39	3.08	35.4
			57.3	**70**	**24.5**	**–**	**115**	**–**	**5.4**	**–**	**–**	**–**	**30.2**
						Eureka Creek							
11	06-13-2008	1245	87	106	46.3	0.75	264	<0.12	20.6	1.06	4.70	0.93	105
	07-14-2008	1150	128	154	64.4	1.01	367	<0.12	28.8	0.93	5.46	1.27	153
	09-09-2008	1510	127	155	68.5	0.83	387	<0.12	30.4	0.98	5.72	1.45	171
	06-03-2009	1655	111	144	57.7	1.56	319	E0.07	26.0	1.24	5.40	1.11	135
	07-13-2009	1900	158	205	80.9	2.08	468	E0.06	36.5	1.15	6.02	1.57	202
	09-08-2009	1500	171	222	86.7	1.76	485	E0.08	40.7	1.13	6.39	1.61	218
	05-31-2010	1930	109	142	63.3	1.36	310	E0.07	27.4	1.16	5.12	1.10	137
	07-12-2010	1930	120	156	66.6	1.00	353	E0.08	30.6	0.95	5.88	1.30	150
	09-06-2010	1805	118	153	71.9	0.83	398	E0.05	31.7	0.97	6.25	1.45	181
	06-06-2011	1910	104	135	57.7	1.04	305	<0.04	26.4	1.27	5.22	1.19	135
	07-20-2011	1900	134	174	70.6	1.18	369	<0.04	32.0	0.94	6.53	1.41	163
	09-05-2011	1755	124	161	65.5	0.82	384	0.04	29.7	0.89	6.40	1.36	163
			124	**159**	**66.7**	**–**	**367**	**–**	**30.1**	**–**	**–**	**–**	**159**

Table 4. Concentrations of major ions and dissolved solids in water samples collected in Kantishna Hills area, Alaska, 2008–11.—Continued

[Number below constituent in parentheses is used by U.S. Environmental Protection Agency and U.S. Geological Survey to identify parameters in computer databases. All values in milligrams per liter. **Site No.:** Site locations are shown in figure 2. **Bold** value indicates average. **Abbreviations:** E, estimated value; <, less than reporting level; –, no data or constitutent not measured]

Site No.	Date	Time	Alkalinity (39086)	Bicarbonate (00453)	Calcium (00915)	Chloride (00940)	Total dissolved solids (70300)	Fluoride (00950)	Magnesium (00925)	Potassium (00935)	Silica (00955)	Sodium (00930)	Sulfate (00945)
						Slate Creek							
12	06-12-2008	1600	64	78	21.9	<0.12	130	<0.12	11.2	0.47	5.13	0.71	41.1
	07-13-2008	1231	79	96	28.0	<0.12	180	E0.07	17.2	0.23	5.08	0.77	62.4
	09-11-2008	0935	92	112	32.4	E0.11	209	E0.07	21.1	0.37	5.19	0.91	76.4
	06-02-2009	1850	81	105	26.1	0.15	150	0.12	17.4	0.54	4.36	0.54	50.3
	07-28-2009	1750	89	116	34.6	<0.12	201	E0.04	18.6	0.33	6.25	0.83	71.5
	09-09-2009	1050	96	125	33.3	E0.11	212	E0.08	19.7	0.38	5.69	0.84	72.2
	06-01-2010	1530	84	109	31.9	0.14	214	0.13	20.0	0.61	4.73	0.68	66.6
	07-13-2010	1515	60	78	24.7	<0.12	145	0.12	15.0	0.22	5.27	0.84	51.4
	09-07-2010	1509	88	114	33.6	0.13	218	E0.06	21.7	0.40	5.62	0.88	78.6
	06-07-2011	1520	73	95	25.6	0.1	159	0.05	16.7	0.50	4.45	0.56	55.0
	07-21-2011	1405	89	116	33.4	<0.06	205	0.06	21.5	0.29	5.74	0.80	81.3
	09-06-2011	1345	99	129	34.1	0.15	237	0.07	23.3	0.36	5.63	0.78	85.8
			83	**106**	**30**	**–**	**188**	**–**	**18.6**	**–**	**–**	**–**	**66.1**
						Eldorado Creek							
13	06-13-2008	1030	136	167	56.5	7.99	293	E0.06	20.1	0.91	6.35	3.57	82.6
	07-14-2008	1530	156	191	65.5	9.56	313	E0.06	21.8	0.78	7.05	4.12	91.8
	09-09-2008	1400	145	177	61.9	6.83	304	E0.06	21.4	0.79	7.33	3.67	96.6
	06-01-2010	0900	161	209	74.5	14.6	330	E0.07	24.8	1.03	7.40	5.44	86.1
	07-12-2010	2120	158	205	69.4	10.0	289	0.10	23.0	0.79	7.51	4.50	83.1
	09-06-2010	1700	130	169	63.1	6.64	302	E0.06	21.8	0.76	8.02	3.45	96.4
			148	**186**	**65.2**	**–**	**305**	**–**	**22.2**	**–**	**–**	**–**	**89.4**
						Friday Creek							
14	06-13-2008	1430	120	147	58.5	0.69	323	<0.12	21.0	1.84	5.12	0.92	107
	07-14-2008	1325	154	188	75.0	0.72	370	<0.12	26.8	1.77	5.54	1.06	143
	09-09-2008	1740	153	187	75.8	0.65	396	<0.12	29.5	1.83	5.46	1.18	155
	06-03-2009	1900	142	185	70.5	1.08	331	E0.08	24.3	1.97	5.70	1.05	127
	07-13-2009	1810	177	230	88.4	1.49	451	<0.08	31.7	2.13	5.84	1.40	175
	09-08-2009	1610	194	252	96.4	1.24	475	E0.05	34.7	2.13	6.36	1.39	192
			157	**198**	**77.4**	**–**	**391**	**–**	**28.0**	**–**	**–**	**–**	**150**
						Moose Creek							
15	06-11-2008	1750	62	76	28.6	1.39	156	E0.06	7.7	0.65	7.15	2.35	40.2
	07-14-2008	0908	76	94	32.1	1.73	156	E0.07	8.5	0.63	8.84	3.18	43.1
	09-08-2008	1755	80	98	36.0	1.59	166	<0.12	9.7	0.69	8.53	3.10	52.9
	06-02-2009	0905	73	95	32.5	2.17	148	E0.09	7.9	0.72	7.76	3.01	40.6
	07-14-2009	1715	85	110	35.8	2.84	166	<0.08	8.7	0.68	8.62	3.41	45.7
	09-08-2009	1750	84	109	35.3	2.20	170	E0.07	8.8	0.68	10.2	3.89	45.1
	05-31-2010	1730	78	101	39.5	3.27	172	0.08	10.7	0.70	7.98	3.06	52.5
	07-12-2010	1720	75	98	33.9	1.89	149	0.09	8.6	0.65	9.33	3.20	41.3
	09-06-2010	1545	72	94	36.4	1.53	174	E0.08	9.7	0.63	9.63	3.39	53.4
	06-06-2011	1730	76	99	36.7	2.21	170	0.08	10.3	0.75	8.01	2.85	52.1
	07-20-2011	2045	78	101	33.2	2.08	156	0.05	9.1	0.67	10.2	3.70	42.7
	09-06-2011	0735	82	107	35.9	1.62	182	0.06	11.0	0.66	9.60	3.15	57.7
			77	**99**	**34.6**	**–**	**164**	**–**	**9.2**	**–**	**–**	**–**	**47.3**

Major ion data also were collected in the 1990s (appendix F) but not in the 1980s. Comparing average values of the concentrations for the 1990s and the current study indicates a slight or no increase in bicarbonate anion. For the sulfate anion, average concentrations were either the same or higher in the samples collected during the present study at Rock Creek, Caribou Creek site 6, Eureka Creek, and Friday Creek, but lower at Glen Creek site 9 and Moose Creek site 15. Average concentrations of calcium and magnesium were higher in the current study at all sites except for Caribou Creek site 3 and Moose Creek site 15.

Trilinear diagrams (Hem, 1985, p. 179) were used to plot the concentrations of dissolved major ions in the water samples. This type of diagram permits the chemical composition of multiple samples to be represented on a single graph, and facilitates classification of the sample chemistry (fig. 23). On the basis of analyses of the samples collected during this study, calcium is the major cation in all streams sampled in Kantishna Hills. Bicarbonate is the predominant anion in the waters of Rock Creek, Moose Creek, Slate Creek, and Eldorado Creek; these waters can be classified as calcium bicarbonate water. Sulfate is the predominant anion at the remaining sites—Caribou Creek (all sites), Glen Creek (all sites), Eureka Creek, and Friday Creek; these waters are classified as calcium sulfate water.

Trace Elements

Samples of the water column, slimy sculpin, and the streambed were collected and analyzed for trace elements. Water-column samples were collected at all sites, sculpin were collected at two sites—Caribou Creek (site 6) and Eureka Creek (site 11)—and streambed sediments were collected at all sites except West Fork Glen Creek (site 8). Water-column samples were analyzed for 23 trace elements in the dissolved phase, sculpin for 22 trace elements, and bed sediments for 34 trace elements.

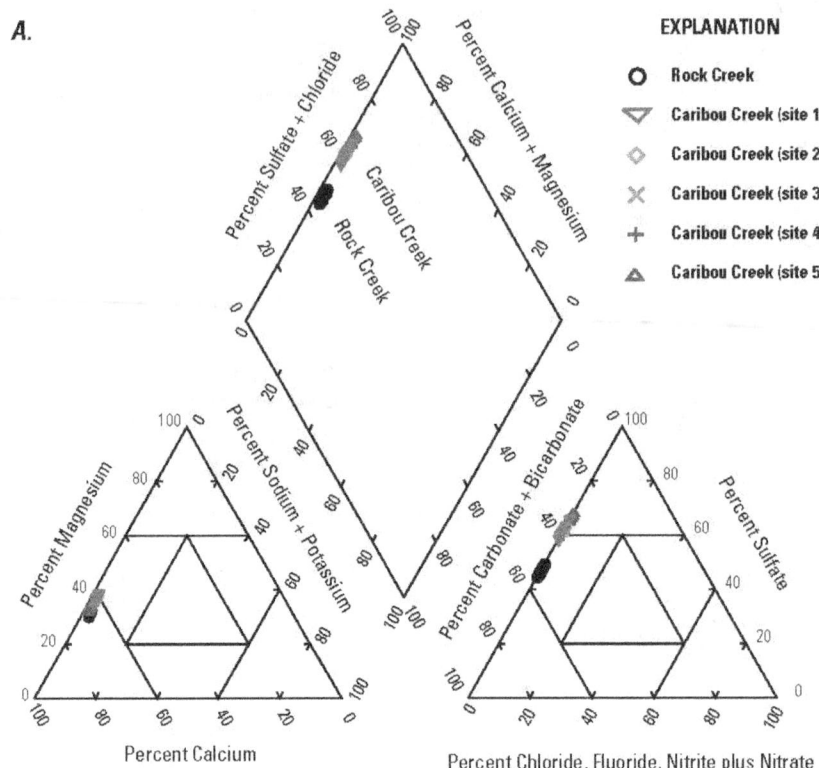

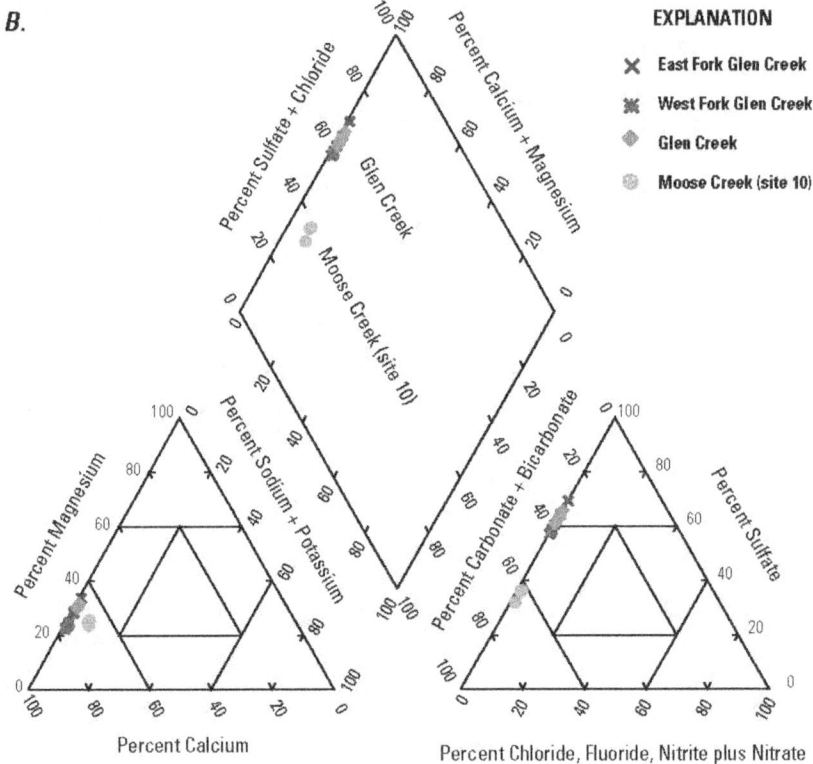

Figure 23. Chemical composition of water samples collected from sites in Kantishna Hills, Alaska.

C.

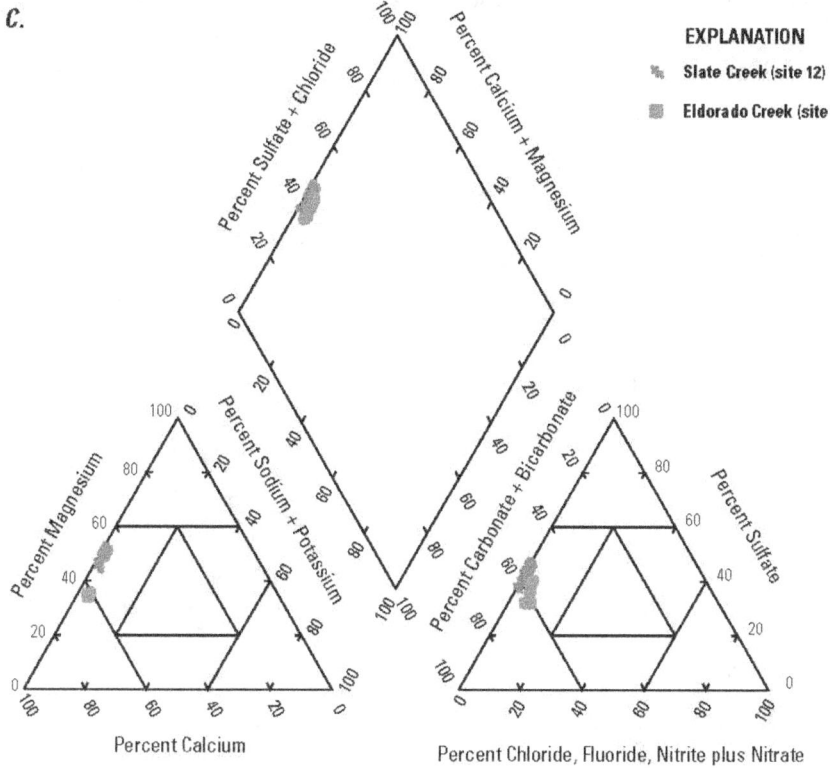

EXPLANATION

Slate Creek (site 12)

Eldorado Creek (site 13)

D.

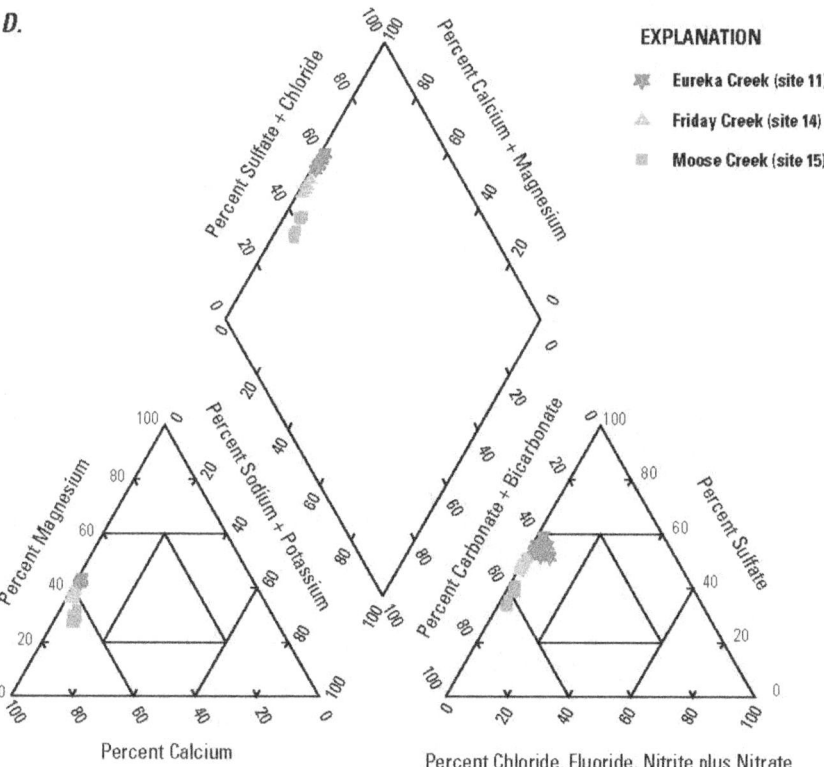

EXPLANATION

Eureka Creek (site 11)

Friday Creek (site 14)

Moose Creek (site 15)

Figure 23.—Continued

For the water-column samples, concentrations of a number of trace elements were less than the respective reporting levels at several sites in 2008 (table 5) and thus were not analyzed in succeeding years. Concentrations of other trace elements collected during 2008–11 were variable both at a particular site and among sites. However, most of the concentrations of trace elements would be considered background levels (Hem, 1985). These trace elements include aluminum, barium, cobalt, copper, lithium, nickel, selenium, and strontium.

Concentrations of the trace elements antimony, arsenic, iron, and manganese were notable at some sites. Typical concentrations of dissolved antimony in natural waters are normally less than 1 μg/L (Hem, 1985). Average concentrations of antimony at Caribou Creek (site 6), Glen Creek, Friday Creek, and Moose Creek (site 15) were slightly above this threshold but average concentrations at Eureka Creek, Slate Creek, and Eldorado Creek were one or two magnitudes higher. Concentrations of arsenic greater than the USEPA drinking water standard (10 μg/L) were exceeded only at Slate Creek. The recommended upper limit for manganese in public water supplies is 50 μg/L; this limit was exceeded at Slate Creek (table 5). For iron in fully aerated streams, concentrations should not exceed more than a few micrograms per liter of dissolved iron in the pH range of about 6.5–8.5 (Hem, 1985). Average concentrations of iron were higher than this threshold at Rock Creek, Caribou Creek (site 6), Eldorado Creek, and Moose Creek (site 15).

At Slate Creek, concentrations of dissolved iron were greater than 1,000 μg/L. As the water that directly drains the antimony deposit mixes with water that drains the non-mined area, pH values rise, and iron oxide from the iron-rich water deposits as flocs and coatings on the alluvial cobbles, which results in the conspicuous red-stained alluvial cobbles along most of the upper reach of the stream (fig. 24).

Table 5.   Dissolved trace-element concentrations from water samples collected at sites located in Kantishna Hills, Alaska, 2008–11.

[Number below constituent in parentheses is used by U.S. Environmental Protection Agency and U.S. Geological Survey to identify parameters in computer databases. All values in micrograms per liter. **Site No.:** Site locations are shown in figure 2. **Bold** indicates average. **Abbreviations:** <, less than; E, estimated; M, measured but not quantified; –, no data or constituent not measured]

Site No.	Date	Time	Alumi-num (01106)	Anti-mony (01095)	Arsenic (01000)	Barium (01005)	Bery-lium (01010)	Boron (01020)	Cad-mium (01025)	Chro-mium (01030)	Cobalt (01035)	Copper (01040)	Iron (01046)	Lead (01049)
							Rock Creek							
1	06-10-2008	1200	17.4	0.52	0.18	12	M	<6	<0.04	E0.06	0.09	1.10	15	E0.07
	07-13-2008	1410	15.3	0.75	0.20	15	E0.01	<6	<0.04	<0.12	0.08	E.86	17	0.18
	09-10-2008	0955	10.6	0.73	0.16	16	<0.01	<6	<0.04	<0.12	0.10	E.55	9	<0.08
	06-03-2009	1420	17.1	0.65	0.20	14	–	–	–	–	0.09	1.20	21	–
	07-28-2009	1620	8.2	1.03	0.21	18	–	–	–	–	0.07	E0.57	9	–
	09-10-2009	1000	9.9	0.86	0.19	18	–	–	–	–	0.13	<1.0	8	–
	06-01-2010	1320	10.6	0.82	0.22	17	–	–	–	–	0.16	E0.57	13	–
	07-13-2010	1355	15.1	0.78	0.19	15	–	–	–	–	0.06	E0.86	24	–
	09-07-2010	1310	12.9	0.75	0.17	16	–	–	–	–	0.07	2.1	10	–
	06-07-2011	1340	15.4	0.65	0.16	13	–	–	–	–	0.05	0.78	20	–
	07-21-2011	1130	8.7	0.89	0.23	16	–	–	–	–	0.04	0.53	10	–
	09-06-2011	1245	13.6	0.73	0.21	16	–	–	–	–	0.06	0.59	8	–
			12.9	**0.76**	**0.19**	**16**	**–**	**–**	**–**	**–**	**0.08**	**1.05**	**14**	**–**
							Caribou Creek							
2	06-12-2008	1120	2.6	0.42	0.52	26	<0.01	<6	E0.02	<0.12	0.04	1	<8	<0.08
	07-15-2008	1019	E1.4	0.51	0.82	33	<0.01	<6	<0.04	<0.12	0.03	<1.0	<8	<0.08
	09-10-2008	1645	E1.3	0.58	0.63	33	<0.01	<6	<0.04	<0.12	0.05	<1.0	<24	<0.08
	06-03-2009	1045	<4.0	0.51	0.61	33	–	–	–	–	0.04	E0.56	<4	–
	07-28-2009	1305	<4.0	0.63	0.86	39	–	–	–	–	0.11	<1.0	<4	–
	09-09-2009	1230	<4.0	0.54	0.76	37	–	–	–	–	0.23	<1.0	<4	–
	06-01-2010	1200	8.9	0.57	0.63	35	–	–	–	–	0.26	E0.62	<4	–
	07-13-2010	1120	<3.4	0.66	0.61	34	–	–	–	–	0.04	<1.0	<6	–
	09-07-2010	1223	E2.1	0.59	0.60	32	–	–	–	–	0.1	E0.60	E5	–
			3.5	**0.56**	**0.67**	**34**	**–**	**–**	**–**	**–**	**0.1**	**0.86**	**7.4**	**–**
3	06-12-2008	940	7.2	1.56	0.93	18	<0.01	<6	<0.04	E0.06	0.04	E.90	E5	<0.08
	07-15-2008	1120	3.7	3.35	1.4	25	<0.01	<6	<0.04	<0.12	0.03	E.63	<8	<0.08
	09-10-2008	1530	3.7	2.85	1.2	25	<0.01	<6	<0.04	<0.12	0.05	<1.0	<8	<0.08
4	06-10-2008	1645	10.8	1.72	0.93	17	<0.01	<6	<0.04	E0.08	0.05	1.1	13	E0.05
	07-15-2008	1240	6.7	3.29	1.1	22	<0.01	<6	<0.04	<0.12	0.03	E.67	E5	<0.08
	09-10-2008	1410	3.7	2.95	1.1	24	<0.01	<6	<0.04	<0.12	0.05	<1.0	<8	<0.08
5	06-10-2008	1345	24.8	1.65	0.78	17	<0.01	<6	<0.04	E0.07	0.23	1.4	86	<0.08
	07-13-2008	1641	32.7	2.88	0.89	20	M	<6	<0.04	<0.12	0.31	1	89	0.13
	09-10-2008	1300	19.4	2.79	0.86	23	<0.01	<6	<0.04	<0.12	0.38	E.92	57	<0.08
6	06-10-2008	1520	27.0	1.65	0.67	17	<0.01	<6	<0.04	E.08	0.16	1.8	76	<0.08
	07-13-2008	1531	37.6	2.73	0.75	20	<0.01	<6	E0.02	E.06	0.21	1.1	82	0.20
	09-10-2008	1150	20.7	2.67	0.74	23	<0.01	<6	<0.04	<0.12	0.29	E.82	43	<0.08
	06-03-2009	1230	31.3	2.07	0.68	19	–	–	–	–	0.20	1.2	86	–
	07-28-2009	1450	25.0	4.71	0.81	26	–	–	–	–	0.22	E0.78	21	–
	09-09-2009	1730	23.2	3.62	0.88	26	–	–	–	–	0.33	E0.54	23	–
	06-01-2010	1420	27.2	2.88	0.62	22	–	–	–	–	0.28	E0.75	54	–
	07-13-2010	1240	42.6	2.98	0.78	22	–	–	–	–	0.22	1.1	107	–
	09-07-2010	1400	30.4	2.61	0.82	21	–	–	–	–	0.21	E0.99	63	–
	06-07-2011	1240	28.9	2.20	0.73	18	–	–	–	–	0.17	1.3	105	–
	07-21-2011	1230	29.8	3.47	0.84	22	–	–	–	–	0.20	0.82	61	–
	09-06-2011	1140	24.3	2.57	0.86	21	–	–	–	–	0.19	0.76	53	–
			29.0	**2.8**	**0.8**	**21**	**–**	**–**	**–**	**–**	**0.22**	**1.15**	**65**	**–**

Table 5.   Dissolved trace-element concentrations from water samples collected at sites located in Kantishna Hills, Alaska, 2008–11.—Continued

[Number below constituent in parentheses is used by U.S. Environmental Protection Agency and U.S. Geological Survey to identify parameters in computer databases. All values in micrograms per liter. **Site No.:** Site locations are shown in figure 2. **Bold** indicates average. **Abbreviations:** <, less than; E,estimated; M, measured but not quantified; –, no data or constituent not measured]

Site No.	Date	Time	Lithium (01130)	Man-ganese (01056)	Moly-bdenum (01060)	Nickel (01065)	Selen-ium (01145)	Silver (01075)	Stron-tium (01080)	Thallium (01057)	Uranium (22703)	Vana-dium (01085)	Zinc (01090)
						Rock Creek							
1	06-10-2008	1200	1.8	4.7	E0.1	1.6	0.41	<0.1	186	<0.04	2.28	E0.03	<1.8
	07-13-2008	1410	2.6	2.9	E0.1	1.4	0.51	<0.1	229	<0.04	2.95	E0.03	<1.8
	09-10-2008	0955	2.5	3.1	E0.1	1.6	0.76	<0.1	255	<0.04	3.19	<0.04	1.9
	06-03-2009	1420	2.3	5.1	–	1.5	0.54	–	221	–	2.92	<0.16	E1.0
	07-28-2009	1620	2.4	1.0	–	1.1	0.55	–	259	–	3.62	<0.16	<2.0
	09-10-2009	1000	2.6	2.2	–	1.4	0.62	–	279	–	4.23	<0.16	<2.0
	06-01-2010	1320	2.6	4.7	–	1.9	0.47	–	257	–	4.18	<0.16	<2.8
	07-13-2010	1355	2.1	2.6	–	1.1	0.42	–	228	–	2.78	<0.16	<2.8
	09-07-2010	1310	2.5	2.3	–	1.2	0.66	–	254	–	3.29	<0.16	E1.8
	06-07-2011	1340	1.7	3	–	1.5	0.53	–	201	–	2.49	<0.08	<1.4
	07-21-2011	1130	2.3	1.7	–	0.9	0.56	–	270	–	3.62	<0.08	<1.4
	09-06-2011	1245	2.2	2.7	–	1.1	0.63	–	259	–	3.19	<0.08	<1.4
			2.3	**3**	–	**1.4**	**0.56**	–	**242**	–	**3.23**	–	–
						Caribou Creek							
2	06-12-2008	1120	1.6	E0.1	<0.2	0.82	0.26	<0.1	480	<0.04	3.13	E0.02	<1.8
	07-15-2008	1019	2.0	E0.1	E0.1	0.52	0.42	<0.1	693	<0.04	4.57	E0.02	<1.8
	09-10-2008	1645	2.0	<0.2	E0.1	0.67	0.46	<0.1	648	<0.04	4.62	E0.02	E1.2
	06-03-2009	1045	1.9	<0.2	–	0.51	0.36	–	621	–	4.35	<0.16	<2.0
	07-28-2009	1305	2.0	<0.2	–	0.72	0.42	–	725	–	5.05	<0.16	E1.3
	09-09-2009	1230	2.5	<0.2	–	1.30	0.39	–	736	–	5.74	<0.16	<2.0
	06-01-2010	1200	1.9	0.9	–	2.50	0.33	–	686	–	5.71	E0.08	E1.4
	07-13-2010	1120	2.1	<0.3	–	0.47	0.27	–	609	–	4.66	<0.16	<2.8
	09-07-2010	1223	2.3	0.9	–	0.44	0.36	–	662	–	5.09	<0.16	<2.8
			2.0	**0.9**	–	**0.88**	**0.36**	–	**651**	–	**4.77**	–	–
3	06-12-2008	940	1.1	0.6	E0.1	1.10	0.30	<0.1	336	<0.04	3.20	E0.03	<1.8
	07-15-2008	1120	1.2	0.3	E0.2	0.74	0.35	<0.1	470	<0.04	4.76	E0.03	<1.8
	09-10-2008	1530	1.3	0.3	E0.1	1.10	0.60	<0.1	451	<0.04	5.16	E0.02	E1.5
4	06-10-2008	1645	E1.0	0.9	E0.1	1.2	0.28	<0.1	288	<0.04	2.93	E0.03	<1.8
	07-15-2008	1240	1.2	0.5	E0.1	0.9	0.35	<0.1	391	<0.04	4.19	E0.03	<1.8
	09-10-2008	1410	1.3	0.8	E0.1	1.1	0.57	<0.1	419	<0.04	4.72	E0.03	E1.1
5	06-10-2008	1345	1.4	10.2	<0.2	1.8	0.26	<0.1	274	<0.04	2.73	0.05	E1.1
	07-13-2008	1641	2.0	11.9	E0.1	1.8	0.30	<0.1	341	<0.04	3.48	0.05	E1.4
	09-10-2008	1300	2.1	15.2	E0.1	2.0	0.52	<0.1	395	<0.04	4.36	E0.03	E1.7
6	06-10-2008	1520	1.6	6.9	<0.2	1.7	0.25	<0.1	272	<0.04	2.55	0.05	E0.95
	07-13-2008	1531	2.1	7.2	E0.1	1.7	0.27	<0.1	320	<0.04	3.19	0.08	E1.0
	09-10-2008	1150	2.1	11.7	E0.1	1.8	0.52	<0.1	388	<0.04	4.17	0.05	E1.7
	06-03-2009	1230	2.1	8.0	–	1.7	0.26	–	308	–	3.03	<0.16	3.5
	07-28-2009	1450	2.8	7.1	–	1.6	0.34	–	387	–	4.41	E0.10	<2.0
	09-09-2009	1730	2.6	11.1	–	2.0	0.40	–	415	–	4.82	<0.16	E1.2
	06-01-2010	1420	2.6	8.5	–	2.8	0.25	–	352	–	4.23	<0.16	E1.7
	07-13-2010	1240	2.1	8.8	–	1.5	0.23	–	314	–	3.18	E0.14	<2.8
	09-07-2010	1400	2.4	9.6	–	1.6	0.35	–	376	–	3.95	E0.09	<2.8
	06-07-2011	1240	2.0	9.7	–	2.3	0.26	–	298	–	2.97	<0.08	<1.4
	07-21-2011	1230	2.3	8.8	–	1.4	0.29	–	366	–	3.89	<0.08	<1.4
	09-06-2011	1140	2.0	9.5	–	1.3	0.34	–	365	–	3.71	<0.08	1.6
			2.2	**8.9**	–	**1.8**	**0.31**	–	**347**	–	**3.68**	–	–

Table 5. Dissolved trace-element concentrations from water samples collected at sites located in Kantishna Hills, Alaska, 2008–11.—Continued

[Number below constituent in parentheses is used by U.S. Environmental Protection Agency and U.S. Geological Survey to identify parameters in computer databases. All values in micrograms per liter. **Site No.:** Site locations are shown in figure 2. **Bold** indicates average. **Abbreviations:** <, less than; E, estimated; M, measured but not quantified; –, no data or constituent not measured]

Site No.	Date	Time	Alumi-num (01106)	Anti-mony (01095)	Arsenic (01000)	Barium (01005)	Bery-lium (01010)	Boron (01020)	Cad-mium (01025)	Chro-mium (01030)	Cobalt (01035)	Copper (01040)	Iron (01046)	Lead (01049)
							East Fork Glen Creek							
7	06-12-2008	1445	E1.3	0.61	4.2	18	M	<6	E0.03	<0.12	0.05	<1.0	<8	<0.08
	07-15-2008	1505	2.8	0.84	5.4	25	<0.01	<6	0.05	<0.12	0.04	E0.70	<8	<0.08
	09-11-2008	0930	E1.5	0.78	3.9	31	<0.01	<6	0.05	<0.12	0.07	<1.0	<8	E0.05
							West Fork Glen Creek							
8	06-12-2008	1300	8.9	0.85	2.4	15	<0.01	<6	E0.03	E0.07	0.04	E0.95	E6	E0.04
	07-15-2008	1405	1.9	1.23	4	26	<0.01	<6	0.05	<0.12	0.03	<1.0	E5	<0.08
	09-11-2008	1535	1.8	2.61	3.7	29	<0.01	<6	0.05	<0.12	0.05	<1.0	<8	E.007
							Glen Creek							
9	06-11-2008	1140	9.5	1.05	1.5	20	<0.01	<6	<0.04	E0.07	0.05	1.4	9	<0.08
	07-15-2008	1050	3.6	1.35	1.7	27	<0.01	<6	<0.04	<0.12	0.03	E.64	<8	<0.08
	09-09-2008	1020	1.9	1.32	1.9	29	<0.01	<6	E0.02	<0.12	0.05	<1.0	<8	<0.08
	06-02-2009	1635	6.9	1.32	1.8	27	–	–	–	–	0.05	1.1	E3	–
	07-14-2009	1000	E2.3	1.39	1.4	28	–	–	–	–	0.09	E0.54	<4	–
	09-09-2009	0920	<4.0	1.45	1.8	31	–	–	–	–	0.15	<1.0	<4	–
	06-01-2010	1630	3.6	1.13	1.4	28	–	–	–	–	0.26	E0.90	E5	–
	07-13-2010	1640	4.3	1.39	1.7	28	–	–	–	–	0.05	E0.69	E4	–
	09-07-2010	1555	E2.5	1.33	2.2	28	–	–	–	–	0.04	<1.0	<6	–
	06-07-2011	1420	5.2	1.32	1.8	26	–	–	–	–	0.03	0.82	<3	–
	07-20-2011	1645	2.7	1.51	1.8	28	–	–	–	–	0.03	<0.50	<3	–
	09-05-2011	1600	3.4	1.32	2.2	29	–	–	–	–	0.04	<0.50	<3	–
			4.6	**1.32**	**1.8**	**27**	–	–	–	–	**0.07**	**1.11**	**5**	–
							Moose Creek							
10	06-11-2008	1425	21.1	0.17	0.39	12	E.01	<6	<0.04	0.15	0.04	1.1	28	<0.08
	07-15-2008	1640	4.8	0.14	0.36	15	<0.01	E3	<0.04	E0.11	0.02	<1.0	E8	<0.08
	09-09-2008	1145	2.6	0.17	0.39	16	<0.01	E3	<0.04	E0.11	0.03	<1.0	<16	<0.08
			9.5	**0.16**	**0.38**	**14**	–	–	–	–	**0.03**	–	–	–
							Eureka Creek							
11	06-13-2008	1245	8.1	6.53	2.3	12	<0.01	<6	E0.03	E0.08	0.05	1.5	20	<0.08
	07-14-2008	1150	3.9	10.30	3.2	16	<0.01	<6	E0.04	<0.12	0.03	E.9	E6	<0.08
	09-09-2008	1510	3.6	8.54	3.7	17	<0.01	<6	0.05	<0.012	0.05	1.3	<8	<0.08
	06-03-2009	1655	6.4	9.90	3.1	17	–	–	–	–	0.04	1.9	9	–
	07-13-2009	1900	4.5	28.10	5.9	20	–	–	–	–	0.09	E0.77	E2	–
	09-08-2009	1500	E2.5	17.00	4.5	19	–	–	–	–	0.13	<1.0	<4	–
	05-31-2010	1930	4.2	9.95	3.1	15	–	–	–	–	0.17	E0.92	8	–
	07-12-2010	1930	5.6	11.70	3.2	17	–	–	–	–	0.04	E0.97	E5	–
	09-06-2010	1805	4.3	9.70	3.3	18	–	–	–	–	0.03	E0.69	E5	–
	06-06-2011	1910	7	8.40	2.8	15	–	–	–	–	0.04	1.2	12	–
	07-20-2011	1900	4.7	11.30	3.7	17	–	–	–	–	0.03	0.61	6	–
	09-05-2011	1755	3.8	7.79	3.5	17	–	–	–	–	0.04	<0.50	9	–
			5.1	**11.60**	**3.5**	**17**	–	–	–	–	**0.06**	**1.30**	**7.8**	–

Table 5. Dissolved trace-element concentrations from water samples collected at sites located in Kantishna Hills, Alaska, 2008–11.— Continued

[Number below constituent in parentheses is used by U.S. Environmental Protection Agency and U.S. Geological Survey to identify parameters in computer databases. All values in micrograms per liter. **Site No.:** Site locations are shown in figure 2. **Bold** indicates average. **Abbreviations:** <, less than; E,estimated; M, measured but not quantified; –, no data or constituent not measured]

Site No.	Date	Time	Lithium (01130)	Man-ganese (01056)	Moly-bdenum (01060)	Nickel (01065)	Selen-ium (01145)	Silver (01075)	Stron-tium (01080)	Thallium (01057)	Uranium (22703)	Vana-dium (01085)	Zinc (01090)
						East Fork Glen Creek							
7	06-12-2008	1445	1.1	<0.2	0.2	1.2	0.35	<0.1	4,848	<0.04	1.97	E0.03	E1.5
	07-15-2008	1505	1.3	E0.1	0.3	1.2	0.57	<0.1	648	<0.04	2.80	E0.03	3.1
	09-11-2008	0930	1.5	E0.1	0.3	1.9	0.77	<0.1	784	<0.04	4.81	E0.03	2.9
						West Fork Glen Creek							
8	06-12-2008	1300	<1.0	0.3	E0.1	1.9	0.18	<0.1	195	<0.04	0.72	E0.02	E1.6
	07-15-2008	1405	E0.8	<0.2	E0.2	1.3	0.32	<0.1	373	<0.04	1.53	<0.04	E1.5
	09-11-2008	1535	E0.7	E0.1	E0.2	1.3	0.37	<0.1	403	<0.04	2.63	<0.04	1.6
						Glen Creek							
9	06-11-2008	1140	E.6	0.6	E0.1	1.40	0.20	<0.1	311	<0.04	2.11	E0.03	<1.8
	07-15-2008	1050	E.9	0.2	E0.2	0.82	0.24	<0.1	454	<0.04	3.05	E0.04	<1.8
	09-09-2008	1020	E.9	E.2	E0.2	1.10	0.33	<0.1	495	<0.04	3.32	E0.03	<1.8
	06-02-2009	1635	E0.9	0.6	–	1.10	0.27	–	439	–	3.13	<0.16	<2.0
	07-14-2009	1000	E0.9	E0.2	–	0.66	0.24	–	421	–	3.20	<0.16	<2.0
	09-09-2009	0920	1.0	0.2	–	1.40	0.32	–	492	–	3.95	<0.16	<2.0
	06-01-2010	1630	0.8	5.4	–	2.60	0.27	–	465	–	3.74	0.18	<2.8
	07-13-2010	1640	1.4	1.0	–	0.82	0.23	–	424	–	3.27	<0.16	<2.8
	09-07-2010	1555	1.2	0.4	–	0.84	0.39	–	512	–	3.60	E0.10	<2.8
	06-07-2011	1420	0.8	0.3	–	0.94	0.22	–	423	–	3.05	<0.08	<1.4
	07-20-2011	1645	0.8	0.2	–	0.58	0.23	–	480	–	3.45	<0.08	<1.4
	09-05-2011	1600	1.1	0.3	–	0.77	0.32	–	502	–	3.50	<0.08	<1.4
			0.9	**0.8**	–	**1.09**	**0.27**	–	**452**	–	**3.28**	–	–
						Moose Creek							
10	06-11-2008	1425	2.0	0.9	E0.2	0.74	0.13	<0.1	125	<0.04	0.39	0.21	E.95
	07-15-2008	1640	2.6	0.8	E0.2	0.41	0.16	<0.1	138	<0.04	0.34	0.22	<1.8
	09-09-2008	1145	2.4	0.7	E0.2	0.40	0.18	<0.1	153	<0.04	0.45	0.17	<1.8
			2.3	**0.8**	–	**0.51**	**0.15**	–	**138**	–	–	–	–
						Eureka Creek							
11	06-13-2008	1245	E.9	1.9	<0.2	1.9	0.15	<0.1	254	<0.04	2.34	E0.03	4.1
	07-14-2008	1150	1.4	0.7	<0.2	1.4	0.20	<0.1	176	<0.04	3.69	E0.02	3.5
	09-09-2008	1510	1.4	1.1	<0.2	1.4	0.27	<0.1	400	<0.04	3.99	<0.04	4.6
	06-03-2009	1655	1.5	1.2	–	1.6	0.17	–	337	–	3.24	<0.16	3.4
	07-13-2009	1900	2.4	0.2	–	1.4	0.21	–	469	–	5.86	<0.16	3.2
	09-08-2009	1500	2.2	0.3	–	1.7	0.21	–	449	–	5.47	<0.16	3.8
	05-31-2010	1930	1.5	1.5	–	2.9	0.17	–	340	–	3.97	<0.16	4.0
	07-12-2010	1930	1.4	0.7	–	1.4	0.15	–	359	–	3.50	<0.16	E2.7
	09-06-2010	1805	1.7	1.3	–	1.4	0.22	–	402	–	4.13	<0.16	3.5
	06-06-2011	1910	1.4	2.5	–	1.8	0.14	–	330	–	3.25	<0.08	2.9
	07-20-2011	1900	1.5	1.3	–	1.2	0.23	–	413	–	4.30	<0.08	3.4
	09-05-2011	1755	1.6	2.7	–	1.2	0.23	–	391	–	3.78	<0.08	4.2
			1.6	**1.2**	–	**1.6**	**0.20**	–	**360**	–	**3.96**	–	**3.7**

Table 5.   Dissolved trace-element concentrations from water samples collected at sites located in Kantishna Hills, Alaska, 2008–11.—Continued

[Number below constituent in parentheses is used by U.S. Environmental Protection Agency and U.S. Geological Survey to identify parameters in computer databases. All values in micrograms per liter. **Site No.:** Site locations are shown in figure 2. **Bold** indicates average. **Abbreviations:** <, less than; E,estimated; M, measured but not quantified; –, no data or constituent not measured]

Site No.	Date	Time	Aluminum (01106)	Antimony (01095)	Arsenic (01000)	Barium (01005)	Beryllium (01010)	Boron (01020)	Cadmium (01025)	Chromium (01030)	Cobalt (01035)	Copper (01040)	Iron (01046)	Lead (01049)
							Slate Creek							
12	06-12-2008	1600	6.8	109	3.3	5	M	<6	E0.02	E0.08	0.82	1.4	661	<0.08
	07-13-2008	1231	13.6	179	6.8	6	M	E3	E0.03	<0.12	1.60	1.5	1,050	0.16
	09-11-2008	0935	6.2	134	5.8	7	E0.01	<6	E0.03	<0.12	2.00	1	1,190	<0.08
	06-02-2009	1850	4.4	116	7.7	6	–	–	–	–	0.59	E0.98	323	–
	07-28-2009	1750	12.8	355	7.0	9	–	–	–	–	2.70	1.0	2,420	–
	09-09-2009	1050	9.4	147	8.2	7	–	–	–	–	2.50	1.5	1,700	–
	06-01-2010	1530	27.9	217	37.8	7	–	–	–	–	4.30	1.9	1,550	–
	07-13-2010	1515	44.3	157	20.7	6	–	–	–	–	2.50	2.0	1,240	–
	09-07-2010	1509	18.7	113	36.1	6	–	–	–	–	3.90	2.0	1,580	–
	06-07-2011	1520	14.8	89.1	36.5	6	–	–	–	–	2.00	0.9	1,450	–
	07-21-2011	1405	39.6	158	69.9	8	–	–	–	–	5.20	1.2	2,980	–
	09-06-2011	1345	45.4	141	53.6	8	–	–	–	–	3.70	1.2	2,510	–
			20.3	**160**	**24.5**	**6.75**	**–**	**–**	**–**	**–**	**2.65**	**1.42**	**1,555**	**–**
							Eldorado Creek							
13	06-13-2008	1030	26.6	34.6	1.8	17	E0.01	<6	0.07	E0.11	0.37	8.5	122	<0.08
	07-14-2008	1530	21.7	47.2	1.7	20	E0.01	<6	0.07	<0.12	0.30	6.3	40	<0.08
	09-09-2008	1400	23.3	38.1	1.4	19	<0.01	<6	0.14	<0.12	0.97	7.3	E13	<0.08
	06-01-2010	0900	13.2	26.8	2.7	22	–	–	–	–	0.38	3.2	156	–
	07-12-2010	2120	17.0	78.2	2.1	23	–	–	–	–	0.28	3.1	23	–
	09-06-2010	1700	33.5	57.4	1.6	19	–	–	–	–	1.10	11.8	19	–
			22.6	**47.1**	**1.9**	**20**	**–**	**–**	**–**	**–**	**0.57**	**6.70**	**72**	**–**
							Friday Creek							
14	06-13-2008	1430	7.2	1.12	2.6	23	<0.01	<6	0.22	0.13	0.05	3.1	E8	0.20
	07-14-2008	1325	4.4	1.24	2.5	28	<0.01	<6	0.24	E.06	0.03	1.3	<8	0.12
	09-09-2008	1740	4.3	1.12	2.5	31	<0.01	<6	0.28	E.06	0.05	E0.83	<8	0.12
	06-03-2009	1900	6.4	1.13	2.7	28	–	–	–	–	0.05	2.1	6	–
	07-13-2009	1810	4.5	1.43	2.9	35	–	–	–	–	0.11	1.3	<4	–
	09-08-2009	1610	E2.9	1.45	2.9	35	–	–	–	–	0.15	E0.81	<4	–
			5.4	**1.25**	**2.7**	**30**	**–**	**–**	**–**	**–**	**0.07**	**1.95**	**6.3**	**–**
							Moose Creek							
15	06-11-2008	1750	18.8	2.79	0.79	16	E0.01	<6	E0.03	0.14	0.11	1.6	67	<0.08
	07-14-2008	0908	4.6	3.03	0.62	19	<0.01	<6	<0.04	E0.07	0.06	E0.65	43	<0.08
	09-08-2008	1755	5.3	3.37	0.75	21	<0.01	<6	<0.04	E0.11	0.11	1.1	33	<0.08
	06-02-2009	0905	7.8	3.40	0.77	18	–	–	–	–	0.12	1.5	44	–
	07-14-2009	1715	E3.0	4.15	0.97	24	–	–	–	–	0.09	E0.52	29	–
	09-08-2009	1750	E2.3	3.48	0.74	21	–	–	–	–	0.09	<1.0	17	–
	05-31-2010	1730	6.3	4.43	1.10	21	–	–	–	–	0.15	E0.79	33	–
	07-12-2010	1720	5.6	4.55	0.78	21	M	–	–	–	0.04	E0.74	11	–
	09-06-2010	1545	8	3.45	0.66	19	–	–	–	–	0.08	1.3	17	–
	06-06-2011	1730	6.8	3.90	0.87	20	–	–	–	–	0.06	1.3	24	–
	07-20-2011	2045	5.3	3.11	0.80	21	–	–	–	–	0.05	0.73	12	–
	09-06-2011	0735	6.3	3.02	0.82	22	–	–	–	–	0.08	1.1	17	–
			7.5	**3.56**	**0.81**	**20**	**–**	**–**	**–**	**–**	**0.09**	**1.23**	**29**	**–**

Table 5.   Dissolved trace-element concentrations from water samples collected at sites located in Kantishna Hills, Alaska, 2008–11.—Continued

[Number below constituent in parentheses is used by U.S. Environmental Protection Agency and U.S. Geological Survey to identify parameters in computer databases. All values in micrograms per liter. **Site No.:** Site locations are shown in <u>figure 2</u>. **Bold** indicates average. **Abbreviations:** <, less than; E,estimated; M, measured but not quantified; –, no data or constituent not measured]

Site No.	Date	Time	Lithium (01130)	Man-ganese (01056)	Moly-bdenum (01060)	Nickel (01065)	Selen-ium (01145)	Silver (01075)	Stron-tium (01080)	Thallium (01057)	Uranium (22703)	Vana-dium (01085)	Zinc (01090)
						Slate Creek							
12	06-12-2008	1600	<1.0	75.7	<0.2	3.9	<0.04	<0.1	73.6	<0.04	0.19	E0.02	5.3
	07-13-2008	1231	<1.0	130	<0.2	5.3	<0.04	<0.1	94	<0.04	0.54	<0.04	9.0
	09-11-2008	0935	E0.8	139	<0.2	6.2	<0.04	<0.1	112	<0.04	1.00	<0.04	10.4
	06-02-2009	1850	E0.6	95.7	–	3.5	<0.06	–	84	–	0.54	<0.16	3.7
	07-28-2009	1750	E0.7	225	–	8.6	<0.06	–	116	–	0.33	<0.16	15.9
	09-09-2009	1050	E0.7	171	–	6.9	<0.06	–	106	–	0.75	<0.16	13.6
	06-01-2010	1530	0.7	493	–	10.9	E0.03	–	105	–	0.74	<0.16	19.9
	07-13-2010	1515	0.5	189	–	6.7	E0.02	–	83.2	–	0.44	<0.16	12.8
	09-07-2010	1509	0.9	278	–	9.8	E0.03	–	110	–	1.04	<0.16	17.0
	06-07-2011	1520	0.6	210	–	6.3	0.04	–	81.8	–	0.39	<0.08	9.2
	07-21-2011	1405	0.7	402	–	12.8	<0.03	–	119	–	0.63	<0.08	22.4
	09-06-2011	1345	0.8	309	–	9.4	0.03	–	124	–	1.03	<0.08	17.0
			0.75	**226**	–	**7.5**	**0.04**	–	**101**	–	**0.64**	–	**13.0**
						Eldorado Creek							
13	06-13-2008	1030	1.6	37.8	E0.1	3.6	0.15	<0.1	390	<0.04	1.47	0.05	14.3
	07-14-2008	1530	1.9	29.5	E0.1	2.9	0.16	<0.1	483	<0.04	1.64	E0.03	14.9
	09-09-2008	1400	2.1	52.2	E0.1	4.7	0.22	<0.1	436	<0.04	1.75	E0.03	35.8
	06-01-2010	0900	2.3	53.9	–	4.2	0.19	–	555	–	1.72	E0.13	13.8
	07-12-2010	2120	2.1	32.3	–	2.4	0.16	–	485	–	1.57	<0.16	8.8
	09-06-2010	1700	2.1	70.2	–	5.0	0.19	–	407	–	1.68	<0.16	22.1
			2.0	**46.0**	–	**3.8**	**0.18**	–	**459**	–	**1.64**	–	**18.3**
						Friday Creek							
14	06-13-2008	1430	1.0	0.5	0.2	1.6	0.18	<0.1	291	<0.04	2.08	0.05	19
	07-14-2008	1325	1.1	E0.1	0.2	1.1	0.20	<0.1	374	<0.04	2.65	0.04	17.7
	09-09-2008	1740	1.2	E0.2	0.2	1.1	0.28	<0.1	389	<0.04	2.90	0.04	22.8
	06-03-2009	1900	1.2	0.3	–	1.2	0.20	–	357	–	2.39	<0.16	14.9
	07-13-2009	1810	1.6	E0.2	–	1.2	0.22	–	450	–	3.53	<0.16	12.9
	09-08-2009	1610	1.7	0.2	–	1.5	0.25	–	449	–	3.55	<0.16	14.6
			1.3	**0.25**	–	**1.3**	**0.22**	–	**385**	–	**2.85**	–	**17.0**
						Moose Creek							
15	06-11-2008	1750	1.7	11.7	E0.2	1.20	0.12	<0.1	174	<0.04	0.61	0.16	4.4
	07-14-2008	0908	2.0	7.2	E0.2	0.62	0.13	<0.1	203	<0.04	0.66	0.13	2.3
	09-08-2008	1755	2.2	9.0	E0.2	0.82	0.19	<0.1	225	<0.04	0.87	0.13	4.1
	06-02-2009	0905	2.4	12.1	–	0.81	0.13	–	202	–	0.66	E0.13	7.00
	07-14-2009	1715	2.6	8.1	–	0.58	0.20	–	238	–	0.81	0.19	E1.3
	09-08-2009	1750	3.1	5.6	–	0.70	0.21	–	213	–	0.78	0.23	E1.6
	05-31-2010	1730	2.1	11.2	–	1.40	0.14	–	258	–	1.03	E0.11	3.1
	07-12-2010	1720	2.2	3.9	–	0.46	0.14	–	212	–	0.70	0.21	E1.5
	09-06-2010	1545	3.2	7.0	–	0.70	0.17	–	215	–	0.83	0.18	E1.6
	06-06-2011	1730	1.6	9.2	–	0.71	0.17	–	224	–	0.86	0.11	3.1
	07-20-2011	2045	2.8	5.0	–	0.49	0.17	–	222	–	0.77	0.21	2.5
	09-06-2011	0735	2.5	6.9	–	0.66	0.17	–	245	–	1.01	0.13	2.8
			2.4	**8.1**	–	**0.76**	**0.16**	–	**219**	–	**0.80**	–	**3.7**

Other metals present at low pH values, such as aluminum, arsenic, cobalt, copper, manganese, and nickel, also are precipitated with the iron oxide flocs and coatings upon mixing.

In the Moose Creek watershed, the concentrations of several trace elements (aluminum, antimony, and arsenic) in Glen Creek (site 9) are diluted after entering Moose Creek as noted by the concentrations at Moose Creek site 10. Downstream of site 10, Eldorado, Eureka, and Friday Creeks enter Moose Creek. Concentrations of several trace elements at these sites are high relative to Moose Creek but are diluted as the streams flow into Moose Creek. As expected, many concentrations at Moose Creek site 15 are higher than at Moose Creek site 10.

Comparison of concentrations of dissolved trace elements analyzed from the current study with published or known values from other studies indicate that concentrations of several elements (for example, antimony and arsenic) were higher during 2008–11 (table 6). Most of the higher concentrations were measured in Slate Creek. In comparing the trace element concentrations with known water-quality standards that have been established for the protection of aquatic life (table 7), only concentrations of iron at Slate Creek exceeded the threshold for chronic concentrations. Concentrations of antimony at Slate Creek and Eureka Creek exceed the USEPA drinking water standard of 6 µg/L.

The comparison of trace element concentrations of sculpin at Caribou Creek site 6 and Eureka Creek were either similar at both sites or higher at Eureka Creek (table 8) except for cobalt and selenium, which were lower. Only selenium concentrations in the sculpin appear to be at levels of potential concern. Typical selenium concentrations in fish at background sites are less than 2 µg/g and whole-body selenium concentrations of 4–6 µg/g were estimated as the threshold for reproductive impairment in sensitive species such as salmon (U.S. Department of the Interior, 1998). Selenium concentrations at the Caribou Creek (site 6) and Eureka Creek sites were 5.6 and 7.3 µg/g, respectively.

Some total and dissolved trace element data were collected in 1982 and 1983 from sites in Kantishna Hills (appendixes B and C). Although these sites were not the same sites sampled in this study, some general observations can be made. In 1982, high concentrations of total antimony were measured in the Caribou Creek watershed (appendix B). Compared to the current study, concentrations of total antimony and arsenic were high in the East Fork of Glen Creek (appendix B, site GN-1) and at the mouth of Glen Creek (appendix B, site GN-2), but were diluted once Glen Creek flowed into Moose Creek (appendix B, site M-1). Concentrations of antimony also were high in Eureka

Figure 24.   Iron or red-stained streambed at site 12 on Slate Creek, Kantishna Hills area, Alaska, August 21, 2008. Channel width is 4.5 feet, average velocity is 1.3 feet per second, and flow is 1.5 cubic feet per second, view looking upstream. The iron stained streambed is the result of iron oxide being deposited. Photograph taken by Dan Long, U.S. Geological Survey.

Creek (0.0248 mg/L) and Friday Creek (0.779 mg/L), but were diluted after they flowed into Moose Creek (appendix B, site M-3). It should be noted that these concentrations were detected from water samples collected downstream of active mining. In 1983, concentrations of many trace elements were below detections limits (appendix C). Samples collected during high flow in Caribou Creek and Friday Creek at sites downstream of active mining had high concentrations of several metals in the total phase. High concentrations of total antimony and total arsenic were measured in the settling pond at Slate Creek and in Eldorado Creek. Most likely, as suspended sediment deposited on the various streambeds, the trace elements remained adsorbed on the sediment.

Table 6. Summary of dissolved trace element concentrations from previous studies compared to ranges of dissolved trace element concentrations from sites in Kantishna Hills, Alaska, 2008–11.

[Elements in **bold** indicate higher concentrations in Kantishna Hills sites. **Abbreviations:** –, not measured; <, less than reporting level; E, estimated value]

Trace element	Martin and Whitfield (1983)	Meybeck (1988)	Hem (1985)	Kantishna Hills (2008–11)
Aluminum	50	40±20	–	1.8–45.4
Antimony	1	–	0.1–1	0.14–355
Arsenic	1.7	1±0.5	0.1–1	0.16–69.9
Barium	60	–	10	5–39
Beryllium	–	–	0.1	<0.01–E0.01
Boron	18	30±20	–	<6–E3
Cadmium	0.02	–	0.1–1	<0.04–0.28
Chromium	1	0.8±0.3	0.1–1	<0.12–0.15
Cobalt	0.2	0.1±0.05	0.1	0.02–5.2
Copper	1.5	2±1	1–10	<0.50–11.8
Iron	40	50±30	–	<3–2,980
Lead	0.1	–	0.1–1	<0.08–0.20
Lithium	12	–	–	0.5–3.2
Manganese	8.2	10±5	–	<0.2–493
Molybdenum	0.5	0.8±0.4	0.1–1	<0.2–0.3
Nickel	0.5	0.4±0.3	0.1–1	0.4–12.8
Selenium	–	–	0.1	<0.03–0.77
Silver	0.3	0.4±0.2	0.1	<0.1
Strontium	60	–	–	73.6–4,848
Vanadium	1	–	1	<0.04–0.23
Zinc	30	10±5	1–10	<1.4–35.8

Table 8. Trace element concentrations measured in fish tissue samples collected from Caribou Creek (site 6) and Eureka Creek (site 11), Kantishna Hills area, Alaska, 2009.

[Parameter code is used by U.S. Environmental Protection Agency and U.S. Geological Survey to identify parameters in computer databases. All values in micrograms per gram. <, less than]

Trace element	Parameter code	Caribou Creek	Eureka Creek
Aluminum	49237	72	280
Antimony	49246	<0.1	0.5
Arsenic	49247	1.1	3.1
Barium	49238	4.2	7.3
Berylium	49248	<0.1	<0.1
Boron	49239	1.6	2.0
Cadmium	49249	0.1	0.3
Chromium	49240	0.9	1.2
Cobalt	49250	0.21	0.12
Copper	49241	2.9	3.0
Iron	49242	200	530
Lead	49251	<0.1	0.6
Manganese	49243	26	25
Mercury	49258	0.094	0.191
Molybdenum	49252	<0.1	<0.1
Nickel	49253	0.76	1.3
Selenium	49254	7.3	5.6
Silver	49255	<0.1	<0.1
Strontium	49244	43	80
Uranium	49257	<0.1	<0.1
Vanadium	49465	0.31	0.91
Zinc	49245	130	230

Table 7. Dissolved concentrations of selected trace elements with established water-quality standards for protection of aquatic life.

[Values from Smith and Huyck (1999) and are in micrograms per liter. **Acute:** Acute toxicity is exposure of toxic substances for a short duration and the adverse effect is high. **Chronic:** Chronic toxicity is exposure of toxic substances for a long duration and the adverse effect is low. –, not reported]

Trace element	Acute	Chronic
Aluminum	750	87
Antimony	9,000	1,600
Arsenic	360	1,90
Beryllium	130	5.3
Cadmium	3.9	1.1
Chromium (VI)	16	11
Copper	18	12
Iron	–	1,000
Lead	82	3.2
Nickel	1,400	160
Selenium	20	5
Silver	4.1	0.12
Thallium	1,400	40
Zinc	120	110

Streambed sediment samples collected at 14 stream sites in the Kantishna Hills area were analyzed for 34 trace elements, 6 major elements, and carbon content (table 9). The most notable concentrations were the high antimony (5,800 and 490 μg/g) and arsenic (3,900 and 690 μg/g) concentrations at Slate and Eldorado Creeks, respectively. Other notable concentrations were high copper (750 μg/g) and manganese (2,400 μg/g) concentrations at Eldorado Creek and the high zinc (870 μg/g) concentrations at Eldorado and Friday Creeks. The highest percentages of aluminum and iron were detected at Slate Creek. Although the geology is different, to provide a general comparison, the concentrations of trace elements in samples from the Kantishna Hill sites were compared to streambed sediments collected from the south side of DENA in 1999 and 2000 (Brabets and Whitman, 2002) (table 10). This general comparison showed about 20 elements with similar ranges in concentrations on the south side of DENA and the Kantishna Hills. Concentrations of trace elements associated with mining such as antimony and arsenic were higher in Kantishna Hills than concentrations from the south side of DENA.

Table 9.   Concentrations of trace elements and percentages of carbon measured in streambed sediments collected at sites in the Kantishna Hills, Alaska, 2008.

[Number below constituent in parentheses is used by U.S. Environmental Protection Agency and U.S. Geological Survey to identify parameters in computer databases. **Site No.:** Locations of sites shown in figure 2. Values in micrograms per gram unless otherwise noted. **Abbreviations:** M, measured but not quantified; <, less than]

Site No.	Site name	Date	Time	Aluminum (percent) (65170)	Antimony (34795)	Arsenic (34800)	Barium (34805)	Berylium (34810)	Bismuth (34816)
1	Rock Creek	06-10-2008	1130	9.2	9.6	20	1,100	2.7	M
2	Caribou Creek	06-12-2008	1045	9.5	12	230	1,500	2.2	M
3	Caribou Creek	06-12-2008	0915	7.7	28	73	830	1.8	M
4	Caribou Creek	06-10-2008	1615	6.9	28	75	600	1.7	M
5	Caribou Creek	06-10-2008	1305	7.2	22	72	680	1.9	M
6	Caribou Creek	06-10-2008	1445	9.8	23	70	1,300	2.6	M
7	East Fork Glen Creek	06-12-2008	1415	8.4	19	210	970	2.3	M
9	Glen Creek	06-11-2008	1115	9.3	16	90	870	2.2	M
10	Moose Creek	06-11-2008	1345	7.8	1.3	6.8	780	1.5	M
11	Eureka Creek	06-13-2008	1200	7.9	72	380	830	2.3	M
12	Slate Creek	08-21-2008	1545	12.0	5,800	3,900	1,300	3.7	M
13	Eldorado Creek	06-13-2008	1000	7.8	490	690	780	2.9	M
14	Friday Creek	06-13-2008	1405	6.6	11	120	720	1.9	M
15	Moose Creek	06-11-2008	1700	7.8	4.7	20	800	1.6	M

Site No.	Site name	Date	Time	Cadmium (34825)	Cerium (34835)	Cesium (65172)	Chromium (34840)	Cobalt (34845)	Copper (34850)
1	Rock Creek	06-10-2008	1130	0.5	200	5.6	82	21	41
2	Caribou Creek	06-12-2008	1045	1.8	120	7.2	100	24	50
3	Caribou Creek	06-12-2008	0915	0.6	140	3.7	84	21	41
4	Caribou Creek	06-10-2008	1615	0.7	120	3.3	74	19	38
5	Caribou Creek	06-10-2008	1305	0.7	150	3.7	78	23	42
6	Caribou Creek	06-10-2008	1445	0.7	250	9.2	100	27	71
7	East Fork Glen Creek	06-12-2008	1415	3.7	91	8.7	260	32	67
9	Glen Creek	06-11-2008	1115	0.8	140	3.7	110	22	49
10	Moose Creek	06-11-2008	1345	0.4	50	3.4	58	12	21
11	Eureka Creek	06-13-2008	1200	4.2	140	8.6	120	30	55
12	Slate Creek	08-21-2008	1545	0.6	160	7.2	120	18	120
13	Eldorado Creek	06-13-2008	1000	5.2	140	6.0	100	33	750
14	Friday Creek	06-13-2008	1405	7.6	120	4.7	160	28	79
15	Moose Creek	06-11-2008	1700	0.6	58	3.5	58	12	25

Site No.	Site name	Date	Time	Gallium (34860)	Iron (percent) (65173)	Lanthanum (34885)	Lead (34890)	Lithium (34895)	Manganese (34905)
1	Rock Creek	06-10-2008	1130	22	4.8	120	36	39	1,100
2	Caribou Creek	06-12-2008	1045	22	5.0	64	120	43	1,000
3	Caribou Creek	06-12-2008	0915	17	4.5	75	47	30	870
4	Caribou Creek	06-10-2008	1615	15	4.3	66	33	27	920
5	Caribou Creek	06-10-2008	1305	16	4.9	88	32	28	1,100
6	Caribou Creek	06-10-2008	1445	25	5.8	140	62	45	650
7	East Fork Glen Creek	06-12-2008	1415	20	6.0	49	50	45	1,100
9	Glen Creek	06-11-2008	1115	20	4.8	77	62	30	1,000
10	Moose Creek	06-11-2008	1345	15	3.4	29	12	46	660
11	Eureka Creek	06-13-2008	1200	22	6.6	80	140	49	1,900
12	Slate Creek	08-21-2008	1545	22	12.0	93	100	35	580
13	Eldorado Creek	06-13-2008	1000	18	8.6	120	77	38	2,400
14	Friday Creek	06-13-2008	1405	20	5.0	64	180	30	940
15	Moose Creek	06-11-2008	1700	16	3.7	31	16	48	690

Table 9. Concentrations of trace elements and percentages of carbon measured in streambed sediments collected at sites in the Kantishna Hills, Alaska, 2008.—Continued

[Number below constituent in parentheses is used by U.S. Environmental Protection Agency and U.S. Geological Survey to identify parameters in computer databases. **Site No.:** Locations of sites shown in figure 2. Values in micrograms per gram unless otherwise noted. **Abbreviations:** M, measured but not quantified; <, less than]

Site No.	Site name	Date	Time	Mercury (34910)	Molybdenum (34915)	Nickel (34925)	Niobium (34930)	Rubidium (65177)	Scandium (34945)
1	Rock Creek	06-10-2008	1130	0.10	0.9	44	23	138	16
2	Caribou Creek	06-12-2008	1045	0.03	1.8	49	18	132	21
3	Caribou Creek	06-12-2008	0915	0.05	1.1	43	20	92.4	16
4	Caribou Creek	06-10-2008	1615	0.05	1.0	42	18	77.5	15
5	Caribou Creek	06-10-2008	1305	0.06	0.9	46	18	86	15
6	Caribou Creek	06-10-2008	1445	0.09	1.3	59	26	176	20
7	East Fork Glen Creek	06-12-2008	1415	0.05	1.6	150	26	113	23
9	Glen Creek	06-11-2008	1115	0.10	0.9	55	20	108	19
10	Moose Creek	06-11-2008	1345	0.07	0.8	22	15	61.6	12
11	Eureka Creek	06-13-2008	1200	0.16	1.2	74	23	122	17
12	Slate Creek	08-21-2008	1545	<0.01	0.6	30	15	185	20
13	Eldorado Creek	06-13-2008	1000	0.16	1.1	76	14	113	15
14	Friday Creek	06-13-2008	1405	0.18	1.6	100	36	103	17
15	Moose Creek	06-11-2008	1700	0.06	1.2	24	16	65.7	13

Site No.	Site name	Date	Time	Selenium (34950)	Silver (34955)	Strontium (34965)	Thallium (04064)	Thorium (34980)	Titanium (percent) (65179)
1	Rock Creek	06-10-2008	1130	1.3	<1.0	140	M	32	0.44
2	Caribou Creek	06-12-2008	1045	0.7	1.5	97	1	22	0.35
3	Caribou Creek	06-12-2008	0915	0.6	<1.0	120	M	25	0.46
4	Caribou Creek	06-10-2008	1615	0.7	<1.0	140	M	22	0.44
5	Caribou Creek	06-10-2008	1305	0.9	<1.0	150	M	23	0.42
6	Caribou Creek	06-10-2008	1445	0.9	<1.0	140	M	46	0.43
7	EF Glen Creek	06-12-2008	1415	1.1	<1.0	160	M	14	0.68
9	Glen Creek	06-11-2008	1115	0.5	<1.0	130	M	22	0.50
10	Moose Creek	06-11-2008	1345	0.3	<1.0	320	M	8	0.49
11	Eureka Creek	06-13-2008	1200	0.7	1.0	180	M	24	0.46
12	Slate Creek	08-21-2008	1545	0.2	<1.0	150	2	21	0.26
13	Eldorado Creek	06-13-2008	1000	0.8	<1.0	200	M	16	0.25
14	Friday Creek	06-13-2008	1405	0.9	3.5	220	M	16	0.63
15	Moose Creek	06-11-2008	1700	0.3	<1.0	330	M	9	0.50

Site No.	Site name	Date	Time	Uranium (35000)	Vanadium (35005)	Yttrium (35010)	Zinc (35020)
1	Rock Creek	06-10-2008	1130	5.7	87	35	150
2	Caribou Creek	06-12-2008	1045	3.1	120	19	240
3	Caribou Creek	06-12-2008	0915	4.0	88	25	120
4	Caribou Creek	06-10-2008	1615	3.5	83	25	120
5	Caribou Creek	06-10-2008	1305	4.5	87	34	160
6	Caribou Creek	06-10-2008	1445	7.9	110	42	180
7	East Fork Glen Creek	06-12-2008	1415	2.4	160	24	330
9	Glen Creek	06-11-2008	1115	3.6	110	30	150
10	Moose Creek	06-11-2008	1345	2.3	97	22	100
11	Eureka Creek	06-13-2008	1200	4.2	100	27	460
12	Slate Creek	08-21-2008	1545	12	91	54	120
13	Eldorado Creek	06-13-2008	1000	5.5	93	69	870
14	Friday Creek	06-13-2008	1405	2.8	120	25	870
15	Moose Creek	06-11-2008	1700	2.5	99	25	130

Table 9.   Concentrations of trace elements and percentages of carbon measured in streambed sediments collected at sites in the Kantishna Hills, Alaska, 2008.—Continued

[Number below constituent in parentheses is used by U.S. Environmental Protection Agency and U.S. Geological Survey to identify parameters in computer databases. **Site No.:** Locations of sites shown in <u>figure 2</u>. Values in micrograms per gram unless otherwise noted. **Abbreviations:** M, measured but not quantified; <, less than]

Site No.	Site name	Date	Time	Calcium (65171)	Magnesium (65174)	Phosphorus (65175)	Potassium (65176)	Sulfur (34970)	Sodium (65178)
1	Rock Creek	06-10-2008	1130	7,700	9,300	1,200	28,000	0.08	10,000
2	Caribou Creek	06-12-2008	1045	9,100	15,000	960	30,000	0.08	11,000
3	Caribou Creek	06-12-2008	0915	10,000	10,000	1,300	20,000	0.13	9,500
4	Caribou Creek	06-10-2008	1615	14,000	9,900	1,200	17,000	0.17	10,000
5	Caribou Creek	06-10-2008	1305	12,000	10,000	1,100	18,000	0.12	10,000
6	Caribou Creek	06-10-2008	1445	6,400	13,000	740	30,000	0.06	7,300
7	East Fork Glen Creek	06-12-2008	1415	28,000	32,000	1,900	24,000	0.08	7,100
9	Glen Creek	06-11-2008	1115	13,000	15,000	1,400	25,000	0.07	13,000
10	Moose Creek	06-11-2008	1345	14,000	8,300	950	14,000	0.02	9,900
11	Eureka Creek	06-13-2008	1200	17,000	11,000	1,200	22,000	0.19	5,700
12	Slate Creek	08-21-2008	1545	2,800	6,300	840	41,000	0.32	2,700
13	Eldorado Creek	06-13-2008	1000	13,000	9,500	1,000	21,000	0.16	5,000
14	Friday Creek	06-13-2008	1405	24,000	18,000	2,100	18,000	0.11	6,100
15	Moose Creek	06-11-2008	1700	14,000	8,500	980	15,000	0.03	9,600

Site No.	Site name	Date	Time	Carbon (total) (percent) (49267)	Inorganic carbon (percent) (49269)	Organic carbon (percent) (49266)
1	Rock Creek	06-10-2008	1130	2.5	0.09	2.4
2	Caribou Creek	06-12-2008	1045	1.6	0.15	1.4
3	Caribou Creek	06-12-2008	915	1.7	0.16	1.5
4	Caribou Creek	06-10-2008	1615	1.5	0.27	1.2
5	Caribou Creek	06-10-2008	1305	2.6	0.17	2.5
6	Caribou Creek	06-10-2008	1445	2.0	0.06	1.9
7	East Fork Glen Creek	06-12-2008	1415	4.0	0.36	3.6
9	Glen Creek	06-11-2008	1115	1.5	0.26	1.3
10	Moose Creek	06-11-2008	1345	1.0	0.21	0.82
11	Eureka Creek	06-13-2008	1200	3.2	0.55	2.7
12	Slate Creek	08-21-2008	1545	2.0	0.14	1.8
13	Eldorado Creek	06-13-2008	1000	4.0	0.42	3.6
14	Friday Creek	06-13-2008	1405	4.2	0.35	3.9
15	Moose Creek	06-11-2008	1700	1.1	0.24	0.86

Table 10. Comparison of ranges of trace element concentrations in streambed sediments from Kantishna Hills and the south side of Denali National Park and Preserve, Alaska.

[Concentrations from Brabets and Whitman (2002). All units in micrograms per gram, except for aluminum, iron, and titanium, in percent. **Abbreviations:** <, less than; M, measured but not quantified]

Area	Aluminum (percent)	Antimony	Arsenic	Barium	Berylium	Bismuth
South Side Denali	5.1–8.5	<0.1–4.2	1.7–88	480–1,400	1.2–4.0	<1.0–1.0
Kantishna Hills	6.6–12	1.3–5,800	6.8–3,900	600–1,500	1.5–3.7	M

	Cadmium	Cerium	Chromium	Cobalt	Copper	
South Side Denali	<0.1–0.5	32–400	3–220	<1–26	3–64	
Kantishna Hills	0.4–7.6	50–250	58–260	12–33	21–750	

	Gallium	Iron (percent)	Lanthanum	Lead	Lithium	Manganese
South Side Denali	13–22	0.4–4.7	16–190	10–76	37–75	240–1,600
Kantishna Hills	15–25	3.4–12	29–140	12–180	27–49	580–2,400

	Mercury	Molybdenum	Nickel	Niobium	Scandium	
South Side Denali	<0.02–0.24	<0.5–1.6	<2–130	7–23	<2–17	
Kantishna Hills	<0.01–0.18	0.6–1.8	22–150	14–36	12–23	

	Selenium	Silver	Strontium	Thallium	Thorium	Titanium (percent)
South Side Denali	<0.1–5.2	<0.1–0.8	69–340	<1–1	6–63	0.1–0.14
Kantishna Hills	0.2–1.3	<1–3.5	97–330	M	8–46	0.25–0.68

	Uranium	Vanadium	Zinc			
South Side Denali	2.4–22	5–170	16–170			
Kantishna Hills	2.3–12	83–160	100–870			

The Canadian Council of Ministers of the Environment (CCME) (1999) established guidelines for the protection of aquatic life for the following eight trace elements in unsieved streambed sediment: arsenic, cadmium, chromium, copper, lead, mercury, nickel, and zinc. Because trace-element samples for this study are from sediments finer than 0.063 mm, where concentrations tend to be greatest, comparisons with the guidelines may overestimate the effects on aquatic organisms (Deacon and Stephens, 1998). However, for this study, it was considered acceptable to compare the concentrations from the finer than 0.063-mm size fraction to the published guidelines.

The CCME guidelines use an assessment value called the probable effect level (PEL), which is the concentration above which adverse effects on aquatic organisms are expected to occur frequently (table 11). MacDonald and others (2000) proposed sediment quality guidelines for eight trace elements, and Van Derveer and Canton (1997) proposed consensus-based guidelines for selenium (table 11). The guidelines of MacDonald and others (2000) and Van Derveer and Canton (1997) used a value called the probable effect concentration (PEC) and assumed a 1 percent organic carbon concentration. The PEC is the concentration above which toxicity is likely. With the exception of mercury and selenium, concentrations were greater than the PEL and the PEC for many of the Kantishna Hills sites (table 11). The concentrations greater than the PEC and PEL most likely reflect the effects of mining rather than the local geology.

Table 11.   Summary of streambed-sediment quality guidelines for nine priority-pollutant trace elements in streambed sediment at selected sites in the Kantishna Hills area, Alaska.

[**Probable Effect Level (PEL):** Values from Canadian Council of Ministers of the Environment (1999). **Probable Effect Concentration (PEC):** Values from McDonald and others (2000). Selenium PEC from Van Derveer and Canton (1997). **Abbreviations:** NG, no guidelines; μg/g, microgram per gram, dry weight]

Constituent	PEL (μg/g)	PEC (μg/g)	Sediment sample concentration greater than PEL	Sediment sample concentration greater than PEC
Arsenic	17	33	All except Moose Creek (site 10)	All except: Rock Creek Moose Creek (sites 10 and 15)
Cadmium	3.53	4.98	East Fork Glen Creek Eureka Creek Eldorado Creek Friday Creek	Eldorado Creek Friday Creek
Chromium	90	111	Caribou Creek (site 2) Caribou Creek (site 6) East Fork Glen Creek Glen Creek Eureka Creek Slate Creek Eldorado Creek Friday Creek	East Fork Glen Creek Eureka Creek Slate Creek Friday Creek
Copper	197	149	Eldorado Creek	Eldorado Creek
Lead	91.3	128	Caribou Creek (site 2) Eureka Creek Slate Creek Friday Creek	Eureka Creek Friday Creek
Mercury	0.486	1.06	None	None
Nickel	36	48.6	All except: Moose Creek (sites 10 and 15) Slate Creek	Caribou Creek (site 2) Caribou Creek (site 6) East Fork Glen Creek Eureka Creek Eldorado Creek Friday Creek Glen Creek
Selenium	NG	4	Not applicable	None
Zinc	315	459	East Fork Glen Creek Eureka Creek Slate Creek Eldorado Creek Friday Creek	Eureka Creek Eldorado Creek Friday Creek

MacDonald and others (2000) also developed a mean PEC quotient, which represents the toxicity of the combined trace element concentrations. The mean PEC quotient is computed by summing the concentrations of all the trace elements normalized by their respective PECs and dividing by the total number of elements. MacDonald and others (2000) determined that sediments with mean PEC quotients less than 0.5 accurately predicted the absence of toxicity in 83 percent of the samples they examined. Mean PEC quotients greater than 0.5 accurately predicted toxicity in 85 percent of the samples.

Comparison of the concentrations of the trace elements with the percentage of organic carbon and mean PEC quotient provides some insights about the bioavailability of these trace elements. The concentration of organic carbon in sediment is an indicator of the amount of organic matter. The capacity of organic matter to concentrate some trace elements in stream sediment is well recognized (Gibbs, 1973; Horowitz, 1991), and this capacity varies with the type of organic matter. For the Kantishna Hills sites, organic carbon concentrations were less than 1 percent only at the two Moose Creek sites (table 12) and the corresponding mean PEC quotients were

less than 0.5. However, the concentrations of the trace elements also were relatively low (table 11). At Rock Creek, the organic carbon content was 2.4 percent, the concentrations of the trace elements were higher than concentrations at the Moose Creek sites, but the mean PEC quotient was less than 0.5. Thus, this site represents an example where high organic carbon content reduces the potential toxicity.

The mean PEC quotient was greater than 0.5 at the remaining sites, which indicated potential toxicity. When normalized to percentage of organic carbon, concentrations of arsenic at nine sites and concentrations of chromium at six sites were greater than the PEC level (table 12). Therefore, concerns of potential toxicity to salmonid eggs and fry could exist where fine-grained sediments are abundant and trace element concentrations exceed PECs. Erosional areas of the streambed, such as riffles, where salmonid eggs typically are deposited and incubated, usually contain low amounts of fine-grained sediments. However, depositional areas such as pools and sloughs, which are preferred rearing habitat for many salmonid fry, are usually more abundant in fine-grained sediments.

Table 12.   Concentrations of priority-pollutant trace elements in streambed sediments finer than 0.063 millimeters in Kantishna Hills, Alaska.

[**Site No.:** Locations of sites shown in figure 2. Concentrations are in micrograms per gram, dry weight and organic carbon is in percent. Values in **bold** are concentrations that when normalized to 1 percent organic carbon, exceed Probable Effect Concentration (PEC), or, for the mean PEC quotient, indicate toxicity for the sum of trace elements, excluding selenium (Macdonald and others, 2000)]

Site No.	Site name	Date	Arsenic	Cadmium	Chromium	Copper	Lead	Mercury	Nickel	Selenium	Zinc	Organic carbon	Mean PEC quotient
1	Rock Creek	06-10-2008	20	0.5	82	41	36	0.10	44	1.3	150	2.4	0.42
2	Caribou Creek 1	06-12-2008	**230**	1.8	100	50	120	0.03	49	0.7	240	1.4	**1.38**
3	Caribou Creek 2	06-12-2008	**73**	0.6	**84**	41	47	0.05	43	0.6	120	1.5	**0.62**
4	Caribou Creek 3	06-10-2008	**75**	0.7	74	38	33	0.05	42	0.7	120	1.2	**0.60**
5	Caribou Creek 4	06-10-2008	72	0.7	**78**	42	32	0.06	46	0.9	160	2.5	**0.61**
6	Caribou Creek 5	06-10-2008	**70**	0.7	**100**	71	62	0.09	59	0.9	180	1.9	**0.73**
7	EF Glen Creek	06-12-2008	**210**	3.7	260	67	50	0.05	150	1.1	330	3.6	**1.77**
9	Glen Creek	06-11-2008	**90**	0.8	**110**	49	62	0.10	55	0.5	150	1.3	**0.78**
10	Moose Creek 1	06-11-2008	6.8	0.4	**58**	21	12	0.07	22	0.3	100	0.82	0.22
11	Eureka Creek	06-13-2008	**380**	4.2	120	55	140	0.16	74	0.7	460	2.7	**2.20**
12	Slate Creek	08-21-2008	**3,900**	0.6	**120**	120	100	<0.01	30	0.2	120	1.8	**15.23**
13	Eldorado Creek	06-13-2008	**690**	5.2	100	750	77	0.16	76	0.8	870	3.6	**4.01**
14	Friday Creek	06-13-2008	120	7.6	160	79	180	0.18	100	0.9	870	3.9	**1.58**
15	Moose Creek 2	06-11-2008	20	0.6	58	25	16	0.06	24	0.3	130	0.86	0.30

Biological Characteristics

Samples for macroinvertebrates and algae were collected at seven sites in the Kantishna Hills area at similar times during the runoff season (August 21–24, 2008, and August 10–12, 2009). Hydrologic conditions for the two sampling years were quite different (table 13). In 2008, sampling took place about a week after a rainy period. Flows were moderately high, ranging from 1.5 to 106 ft³/s, and average streamflow velocities were 1.2–2.6 ft/s. Water temperatures reflected the cool summer and ranged from 2.0–8.5°C. In contrast, the 2009 sampling took place during a relatively warm, dry summer. Discharges and velocities in 2009 were about 80 percent less than the discharges in 2008 (0.2–24.3 ft³/s) and velocities in 2008 were about 50 percent less (0.6–1.2 ft/s). Water temperatures were higher at all sites in 2009; the water temperature at Rock Creek was 10.0°C in 2009 compared to 2.9°C in 2008.

Characteristics of each stream reach were documented with a set of notes and photographs, depicting major geomorphic channel units such as riffles and pools, and other features such as woody debris. Slate Creek (site 12) and Friday Creek (site 14) are small watersheds (drainage areas 0.5 and 1.5 mi², respectively) and the channels are remnants of past mining with little or no vegetation. Caribou Creek (site 2), Glen Creek (site 9), and Eureka Creek (site 11) are mid-sized watersheds (drainage areas 3.8–5.0 mi²) and the channels consist of gravel and small boulders. Rock Creek (site 1) and Caribou Creek (site 6) were the largest watersheds sampled (drainage areas 19.2 and 34.2 mi², respectively). The channel of Rock Creek consists of medium to large boulders and no woody debris was present in the reach sampled. The channel of Caribou Creek consisted of small to medium size cobbles and woody debris was present in a few locations along the sampled reach.

Macroinvertebrates

Macroinvertebrate data from the Kantishna Hills sites were collected using the USGS NAWQA Program sampling protocols (Cuffney and others, 1993). A total of 90 taxa (appendix G) were found at the Kantishna Hills sites during the 2008 and 2009 sampling periods. Eighty-four percent of the taxa identified were insects. Dipterans (flies) comprised 53 percent of taxa, followed by Plecopterans (stoneflies, 13 percent), Ephemeropterans (mayflies, 11 percent), Trichopterans (caddisflies, 8 percent), Arachnids (water mites, 7 percent), Collembolans (springtails, 3 percent), and Gastropods (snails, 2 percent). Turbellarians (flatworms), Annelids (aquatic worms), Ostracods (clam shrimp), and Nematodes (roundworms) each comprised 1 percent.

With respect to all macroinvertebrates collected in Alaska using NAWQA protocols, 14 taxa were found only at the Kantishna Hills sites and insects comprised 86 percent of the these taxa: Ephemeroptera, Plecoptera, and Diptera (appendix G). Oligochaetes (aquatic worms) and the stonefly, *Zapada oregonensis*, were found at all Kantishna Hills sites. The lack of a more specific taxonomic identification of oligochaetes renders the occurrence at all sites a diluted metric, as members of the class Oligochaeta occupy many different habitat types and are found in a variety of water quality conditions. *Zapada oregonensis* is a ubiquitous stonefly in Alaska, and is commonly found in samples from different physical and chemical environments. Both samples collected from Friday Creek and the Eureka Creek 2009 sample showed *Zapada oregonensis* to be the most dominant taxa. Oligochaetes were the second most dominant taxa identified in four samples (table 14).

According to the calculations, the NICRI index ranks Caribou Creek (site 6) as having the best overall condition relative to the macroinvertebrate community, followed by Rock Creek (site 1) (table 15). As noted, no mining has occurred in the Rock Creek watershed; therefore, it is considered to be in an undisturbed condition. Caribou Creek (site 6) most likely scored well because it is at a lower elevation (most downstream site) and has mostly riffles (good macroinvertebrate habitat) as opposed to step-pools. Caribou Creek #1 (site 2) did not score as high, because it is at a much higher altitude and is a much smaller, higher gradient, step-pool stream with considerably less available habitat for macroinvertebrates. Disturbance from past mining upstream of Caribou Creek (site 6) appears to have had minimal effect, probably because of the distance from mining sites and the time since the last major mining projects (mid 1980s). The lowest scoring sites or samples were Slate Creek (site 12) and Friday Creek (site 14). Both of these creeks clearly exhibit the scars of placer mining along their banks in the form of non-vegetated tailing piles consisting of unconsolidated gravels, cobbles and boulders. Slate Creek has severely embedded (surrounded or covered by fine sediments) and rust-colored (high levels of iron) streambeds, resulting in sub-optimal habitats for a diverse macroinvertebrate community (Hynes, 1970; Thorp and Covich, 1991).

The macroinvertebrate data from Kantishna Hills were compared with other macroinvertebrate data in Alaska collected by the USGS using the same data collection protocols. Comparisons to national data were avoided because high latitude temperature conditions affecting macroinvertebrate refugia (for example, bed sediments that freeze more or less under specific conditions) may skew results (Irons and others, 1989; Scrimgeour and others, 1994) when compared to temperate areas in the lower 48 states. The NAWQA Alaska data are from sites located on the south side

Table 13. Hydrologic conditions during macroinvertebrate and algae sampling, Kantishna Hills area, Alaska, 2008–09.

[Site No.: Locations of sites shown in figure 2. **Abbreviations:** ft^3/s, cubic foot per second; ft/s, foot per second; °C, degrees Celsius]

Site No.	USGS station No.	Site name	2008 sampling			2009 sampling		
			Discharge (ft^3/s)	Average velocity (ft/s)	Water temperature (°C)	Discharge (ft^3/s)	Average velocity (ft/s)	Water temperature (°C)
1	15547510	Rock Creek	65.9	2.6	2.9	13.6	1.1	10
2	15547520	Caribou Creek	14.4	2.2	3.8	2.8	1.2	6.1
6	15547530	Caribou Creek	106	2.6	8.5	24.3	1.2	12.5
9	15547560	Glen Creek	19.1	2.1	4.9	2.6	0.6	6.3
11	15547570	Eureka Creek	16.9	2.4	2.4	4.1	1.1	9.3
12	15547575	Slate Creek	1.5	1.2	4.8	0.3	1.1	8.5
14	15547590	Friday Creek	3.1	1.9	2.0	0.2	0.6	3.8

Table 14. Taxa identified as most and second most dominant at selected sites, Kantishna Hills area, Alaska, 2008–09.

Site No.	USGS station No.	Site name	Collection date	Most dominant taxa	Percent dominance	Second most dominant taxa	Percent dominance
1	15547510	Rock Creek	08-23-2008	*Epeorus deceptivus*	19.4	Oligochaeta	15.5
			08-12-2009	*Epeorus deceptivus/hesperus*	27.0	Oligochaeta	14.4
2	15547520	Caribou Creek	08-24-2008	*Baetis bicaudatus*	40.5	Oligochaeta	16.3
			08-11-2009	*Ameletus* sp.	22.7	*Epeorus deceptivus/hesperus*	21.8
6	15547530	Caribou Creek	08-23-2008	*Prosimulium* sp.	23.8	*Epeorus deceptivus*	20.7
			08-11-2009	*Baetis bicaudatus*	20.6	*Taenionema* sp.	11.4
9	15547560	Glen Creek	08-22-2008	*Taenionema* sp.	31.3	*Prosimulium* sp.	15.3
			08-12-2009	*Taenionema* sp.	60.5	*Zapada oregonensis* gr.	8.7
11	15547570	Eureka Creek	08-22-2008	Oligochaeta	29.1	*Zapada oregonensis* gr.	20.6
			08-10-2009	*Zapada oregonensis* gr.	19.7	*Oreogeton* sp.	16.5
12	15547575	Slate Creek	08-21-2008	*Nemoura* sp.	30.8	*Capnia* sp.	29.8
			08-11-2009	*Diamesa* sp.	19.4	*Eukiefferiella* sp.	17.6
14	15547590	Friday Creek	08-24-2008	*Zapada oregonensis* gr.	30.7	Oligochaeta	12.7
			08-11-2009	*Zapada oregonensis* gr.	34.2	*Pagastia* sp.	9.9

of DENA, Wrangell–St. Elias National Park and Preserve (southcentral Alaska), the Cook Inlet watershed (southcentral Alaska), and Lake Clark National Park and Preserve (southcentral, southwest Alaska). Most of the sites represent pristine undeveloped watersheds except for some sites in Anchorage, which represent urban watersheds.

Comparisons of the data were made by visual and statistical methods. Visually, violin plots, which combine the information from a boxplot and a histogram, were used to show the distribution and the structure of the data. The violin plots illustrate a kernel density estimation that represents a smooth fit to a rotated histogram, and provides an estimated distribution of points. The wider the bow of the violin plot, the greater the density of actual and estimated values is within that area of the plot. The beeswarm or dot plot overlaying each violin plot represents the actual values used to calculate the violin plots. The violin plots were created using the R software (R Development Core Team, 2010), the vioplot package (Adler, 2005) and the beeswarm package (Eklund, 2010). Statistically, the non-parametric Anderson-Darling k-sample test ($p < 0.05$) was used to determine if the two data sets per metric were from the same population. It should be noted that only 14 samples were collected for the Kantishna Hills sites, and with additional samples, results of the k-sample test might change.

Table 15. Overall rank of National Invertebrate Community Ranking Index metrics for sites in Kantishna Hills, Alaska, 2008–09.

[Table includes all Alaska sites and Kantishna Hills sites. **Site No.:** Locations of sites shown in figure 2. **Overall rank:** 1 = least disturbed, 14 = most disturbed. **Abbreviation:** EPT, Ephemeroptera, Plecoptera, Trichoptera]

Site No.	USGS station No.	Site name	Collection date	Overall rank	Shannon diversity	Evenness	Abundance of tolerant taxa	Richness of tolerant taxa	Two most dominant taxa (percent)	Taxa richness	EPT richness	Predator richness	Richness of EPT taxa (percent)	Gatherer/ collector richness	Chironomid richness (percent)	Total
1	15547510	Rock Creek	08-23-2008	4	0.685	0.828	0.858	0.9	0.843	0.5	0.642	0.226	0.75	0.822	0.762	7.816
2	15547520	Caribou Creek	08-24-2008	10	0.157	0.328	0.558	0.986	0.229	0.035	0.297	0.047	0.952	1	0.953	5.542
6	15547530	Caribou Creek	08-23-2008	2	0.671	0.6	0.986	0.829	0.615	0.642	0.833	0.416	0.785	0.929	0.881	8.187
9	15547560	Glen Creek	08-22-2008	6	0.357	0.471	0.9	0.772	0.558	0.738	0.833	0.321	0.702	0.822	0.834	7.308
11	15547570	Eureka Creek	08-22-2008	7	0.571	0.714	0.486	0.815	0.429	0.071	0.392	0	0.988	1	1	6.466
12	15547575	Slate Creek	08-21-2008	14	0.185	0.457	0.943	0.186	0.186	0.119	0.023	0.047	0.083	0.893	0.191	3.313
14	15547590	Friday Creek	08-24-2008	11	0.528	0.485	0.286	0.243	0.686	0.309	0.297	0.047	0.619	1	0.905	5.405
1	15547510	Rock Creek	08-12-2009	3	0.728	0.614	0.843	0.8	0.715	0.785	0.952	0.583	0.88	0.608	0.667	8.175
2	15547520	Caribou Creek	08-11-2009	8	0.457	0.571	0.972	0.515	0.6	0.095	0.392	0	0.952	0.929	0.75	6.233
6	15547530	Caribou Creek	08-11-2009	1	1	0.957	0.815	0.858	0.929	0.928	0.988	0.416	0.845	0.893	0.798	9.427
9	15547560	Glen Creek	08-12-2009	9	0.114	0.128	1	0.743	0.115	0.345	0.75	0	0.952	1	0.846	5.993
11	15547570	Eureka Creek	08-10-2009	5	0.814	0.971	0.658	0.786	0.8	0.369	0.642	0.047	0.892	0.893	0.87	7.742
12	15547575	Slate Creek	08-11-2009	13	0.871	1	0.043	0.358	0.815	0.035	0.023	0.154	0.142	0.893	0.12	4.454
14	15547590	Friday Creek	08-11-2009	12	0.585	0.685	0.358	0	0.643	0.166	0.214	0.047	0.63	0.822	0.465	4.615

The violin plots of the Alaska data show a fairly normal distribution for eight of the metrics (fig. 25). Two of the metrics, Shannon diversity (fig. 25I and Evenness (fig. 25J), show a strong negative skew and one metric, Percentage of two most dominant taxa (fig. 25K), shows a strong positive skew. Both the negative and positive skews represent data from a highly urbanized stream in Anchorage and a glacier-fed river that was sampled just after ice breakup, respectively. Samples collected from these two sites had a very low diversity of insects and few generalist taxa. Violin plots of the Kantishna Hills data showed no strong negative or positive skews.

The Anderson-Darling test indicated statistically significant difference between the two populations for six metrics. Two of the metrics, percent Ephememeroptera, Plecoptera, and Trichoptera (EPT))-richness and percent chironomid richness measure the percentage of taxa belonging to the orders Ephemeroptera, Plecoptera, and Trichoptera and the percentage of total richness composed of midges (chironomids). The Kantishna Hills sites had generally higher percentages of EPT taxa, an indication of good water quality, compared to other Alaskan sites. The percent chironomid richness is a measure of the percentage of total richness composed of midges. The Kantishna Hills sites had generally lower values, which may suggest that these sites were in better condition relative to land use than sites comprising the Alaskan data. The Slate Creek and Friday Creek sites had the lowest values for both percent EPT and percent chironomid richness of these metrics, and both sites were associated with the tail of the violin plots (figs. 25C and 25D). These streams were the most severely affected by mining and ranked near the bottom of the NICRI.

Functional group richness metrics determined to be significantly different from the Alaskan data set were predator richness and gatherer/collector richness. Predator richness of the Kantishna Hills sites tended to be much lower overall than Alaskan samples and all but one of the values was less than the median for the Alaska sites. This suggests that either the timing of sample collection was not conducive to a rich assemblage of macroinvertebrate predators, or that this area has a naturally low occurrence of predatory macroinvertebrates. The gatherer/collector richness metric is positively correlated with increasing land use disturbance. Values from Kantishna Hills were low when compared with Alaskan values. Rock Creek samples (R8, R9), followed by Friday and Slate Creeks in 2009 (F9, S9), showed the greatest values, although they were very low overall when compared to the Alaska data. The greatest values were again associated with the tail of the Kantishna Hills violin plot, but all values were considered low.

The two tolerance metrics that indicated significant differences between data sets were abundance of tolerant taxa and richness of tolerant taxa. The abundance of tolerant taxa is an abundance-weighted USEPA tolerance value for the sample and a positive response to increasing disturbance. Slate and Friday Creeks were associated with increased values and therefore poorer conditions relative to the other Kantishna Hills sites. The richness of tolerant taxa is defined as the average USEPA tolerance values for a sample based on richness, and responds positively to perturbation. Similar to the previous metric, Slate and Friday Creeks had by far the greatest values, which is further indication that both these creeks have been substantially disturbed and continue to show some evidence of this disturbance.

Another way to compare the macroinvertebrate communities among sites is through the use of an ordination resulting from a multivariate-statistical analysis, such as non-metric multi-dimensional scaling (NMDS). Macroinvertebrate presence or absence data were standardized to the lowest identifiable taxonomic level for each sample in order to generate a Bray-Curtis dissimilarity (distance) input matrix. For macroinvertebrate presence or absence data from sites sampled in 2008 and 2009 (fig. 26), the first (x) axis provides the greatest discrimination among sites and appears to distinguish the 2009 sites from the 2008 sites. The separation likely is due to the difference in weather between 2008 (wet, cool) and 2009 (dry, warm). The second (y) axis by definition has less discriminatory power than the first axis, and may show the effect watershed size. The two small watersheds (Friday Creek and Slate Creek), the three midsize watersheds (Caribou Creek [site 2], Eureka Creek, and Glen Creek), and the larger watersheds (Caribou Creek [site 6], Rock Creek) tended to group together.

The second NMDS plot (fig. 27) consists of all macroinvertebrate data collected in Alaska by the USGS. Data collected from the Kantishna Hills area and from Wrangell-St. Elias National Park (WRST) form a distinct group compared to the rest of the sites in Alaska sampled by the USGS. The other sites, not located in Kantishna Hills or WRST (table 16), are primarily located on streams in Anchorage (34 sites), and on streams located on the Kenai Peninsula (21 sites), south of Anchorage. These areas and the remaining sites represent a transitional climate between maritime and continental (Brabets and others, 1999). Because most of the sites shown on this plot represent minimally disturbed areas, the distinction among groups likely reflects the effects of Interior Alaska climate.

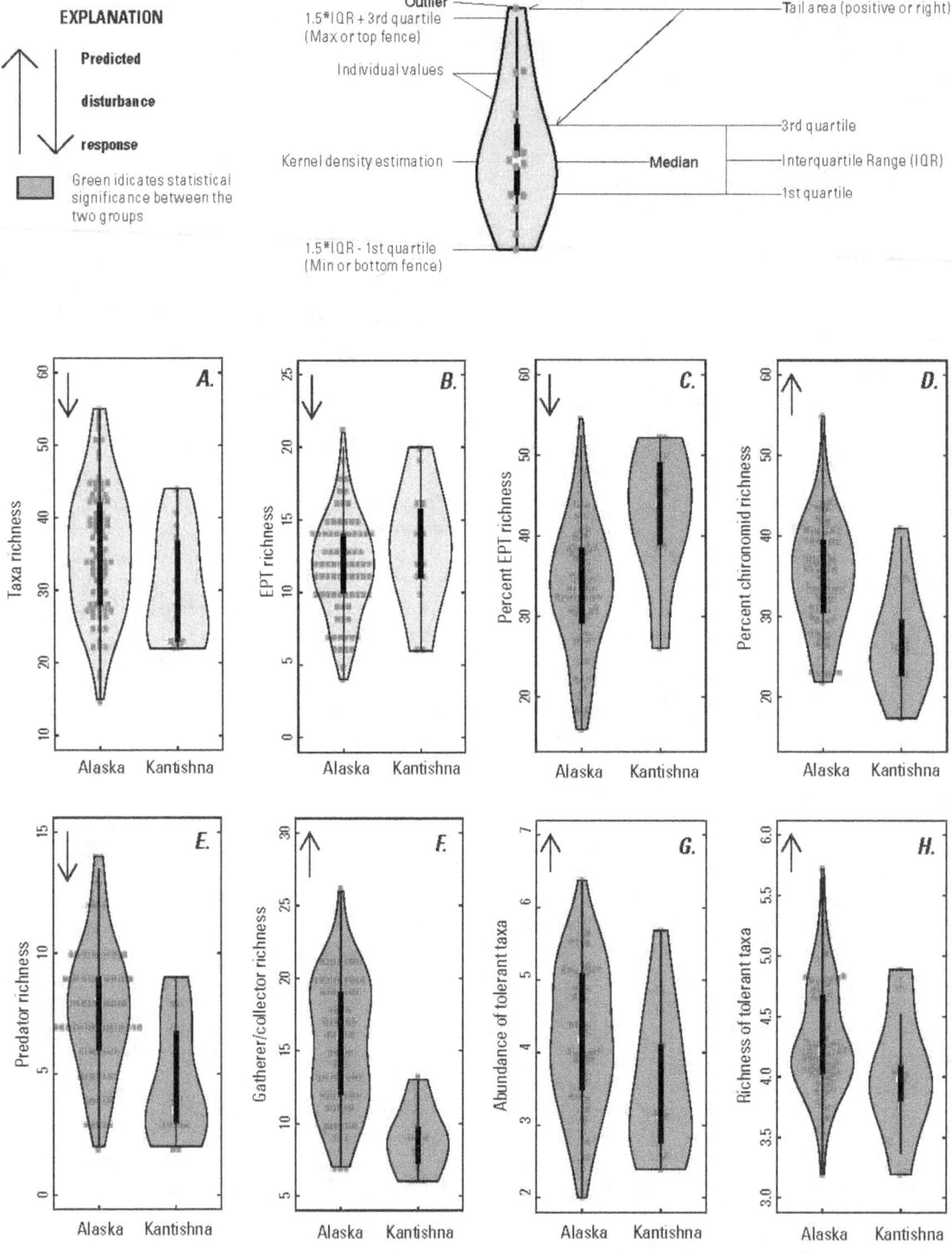

Figure 25. Macroinvertebrate data for 11 metrics for Kantishna Hills sites and other sites in Alaska.

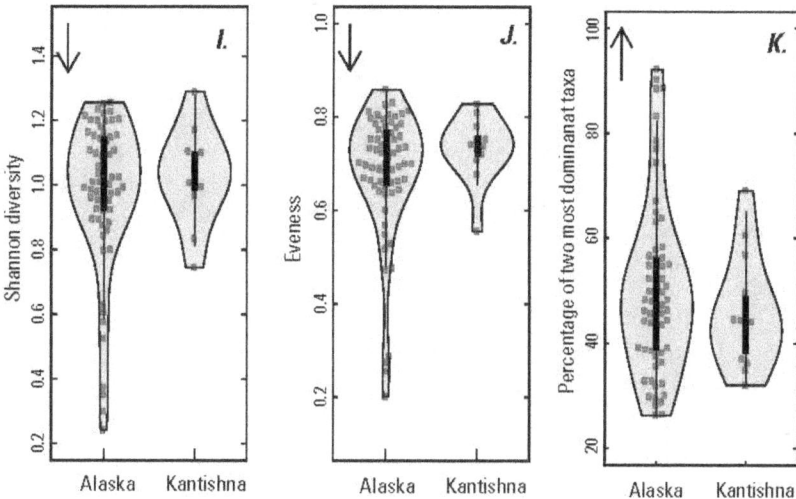

Figure 25.—Continued

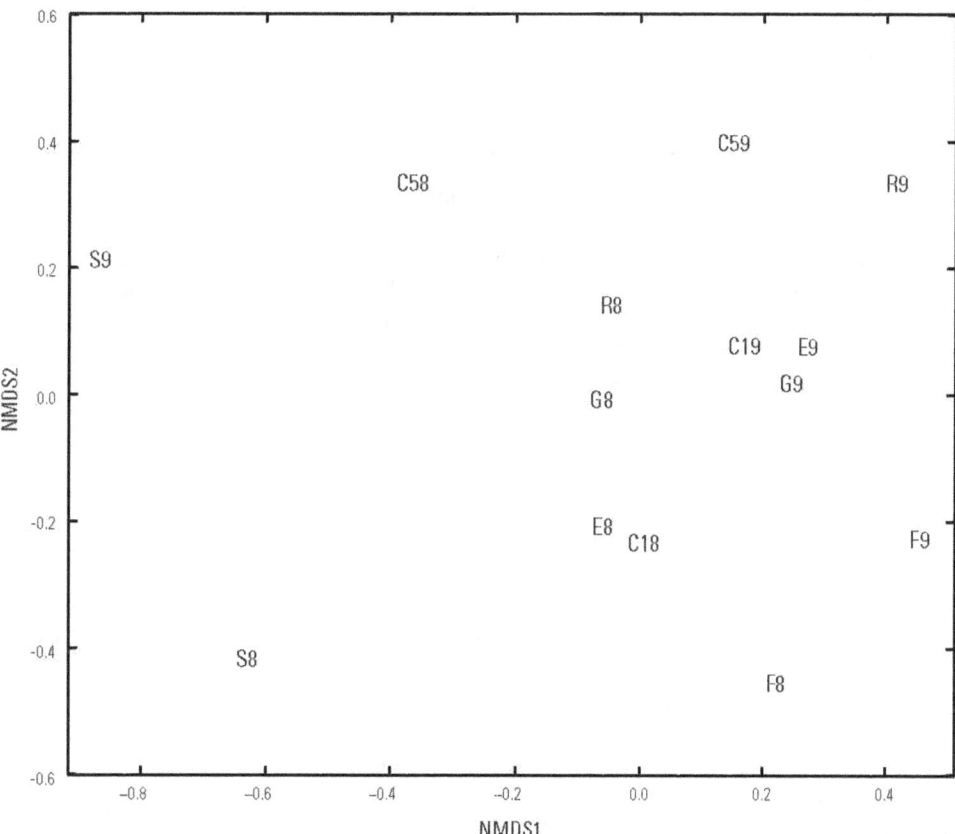

Figure 26. Similarity of macroinvertebrate communities, based on the first and second dimension from a non-metric multi-dimensional scaling (NMDS) analysis, at selected sites in Kantishna Hills, Alaska, 2008–09. Site numbers are given in table 16.

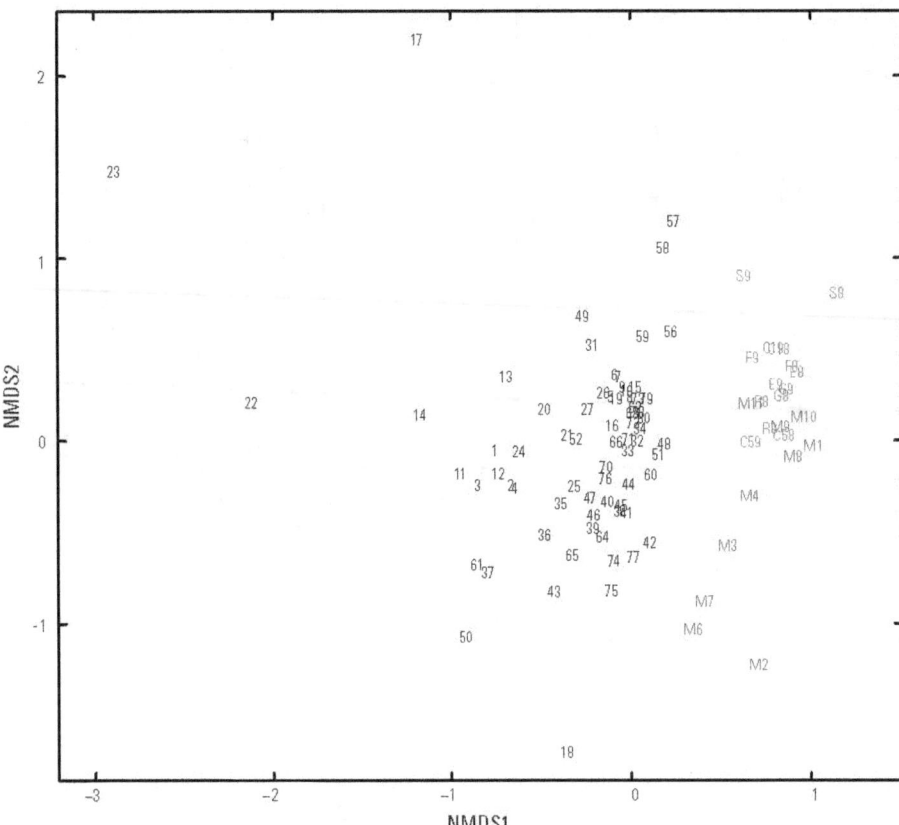

Figure 27.   Similarity of macroinvertebrate communities, based on the first and second dimension from non-metric multi-dimensional scaling (NMDS), at selected sites in Kantishna Hills and other sites sampled by the U.S. Geological Survey, Alaska, 2008–09. Site numbers in color are from Kantishna Hills and Wrangell St. Elias National Park and Preserve. Site numbers are given in table 16.

Table 16.   Stations used in non-metric multi-dimensional scaling analysis, Alaska.

[USGS, U.S. Geological Survey]

Site No.	USGS station No.	Site name
1	15266110	Kenai River below Skilak Lake outlet, near Sterling, Alaska
2	15266300	Kenai River at Soldotna, Alaska
3	15266110	Kenai River below Skilak Lake outlet, near Sterling, Alaska
4	15266300	Kenai River at Soldotna, Alaska
5	15273040	Rabbit Creek at Porcupine Trail Road, near Anchorage, Alaska
6	15273030	Rabbit Creek at East 140 Avenue, near Anchorage, Alaska
7	15273020	Rabbit Creek at Hillside Drive, near Anchorage, Alaska
8	15273100	Little Rabbit Creek near Anchorage, Alaska
9	15273097	Little Rabbit Creek at Goldenview Drive, near Anchorage, Alaska
10	15273090	Little Rabbit Creek at Nickleen Street, near Anchorage, Alaska
11	15266110	Kenai River below Skilak Lake outlet, near Sterling, Alaska
12	15266300	Kenai River at Soldotna, Alaska
13	15266010	Kenai River below Russian River, near Cooper Landing, Alaska
14	15266020	Kenai River at Jim's Landing, near Cooper Landing, Alaska
15	15283550	Moose Creek above Wishbone Hill, near Sutton, Alaska
16	15283700	Moose Creek near Palmer, Alaska
17	15274000	South Fork Campbell Creek near Anchorage, Alaska
18	15274000	South Fork Campbell Creek near Anchorage, Alaska
19	15274000	South Fork Campbell Creek near Anchorage, Alaska
20	15274557	Campbell Creek at Creek Street, near Anchorage, Alaska
21	15274395	Campbell Creek at New Seward Highway, near Anchorage, Alaska
22	15275100	Chester Creek at Arctic Boulevard, at Anchorage, Alaska
23	15275100	Chester Creek at Arctic Boulevard, at Anchorage, Alaska
24	15275100	Chester Creek at Arctic Boulevard, at Anchorage, Alaska
25	15274830	South Branch of South Fork Chester Creek, at Boniface Parkway near Anchorage, Alaska
26	15274796	South Branch of South Fork Chester Creek at Tank Trail, near Anchorage, Alaska
27	15292700	Talkeetna River near Talkeetna, Alaska
M10	15210700	Chokosna River at McCarthy Road, near Chitina, Alaska
M2	15210250	Long Lake Creek Tributary 2 at McCarthy Road, near McCarthy, Alaska
M6	15210300	Long Lake Creek at McCarthy Road, near McCarthy, Alaska
31	585750154101100	Kamishak River near Kamishak, Alaska
32	15274000	South Fork Campbell Creek near Anchorage, Alaska
33	15276570	Ship Creek below Power Plant, at Elmendorf AFB, Alaska
34	15276200	Ship Creek at Glenn Highway, near Anchorage, Alaska
35	15294100	Deshka River near Willow, Alaska
36	15241600	Ninilchik River at Ninilchik, Alaska
37	15294100	Deshka River near Willow, Alaska
38	600204151401800	Deep Creek 0.6 mile above Sterling Highway, near Ninilchik, Alaska
39	600321151325000	Ninilchik River below Tributary 3, near Ninilchik, Alaska
40	15241600	Ninilchik River at Ninilchik, Alaska
41	15239840	Anchor River above Twitter Creek, near Homer, Alaska
42	15240000	Anchor River at Anchor Point, Alaska
43	600945151210900	Ninilchik River 1.5 miles below Tributary1, near Ninilchik, Alaska
44	595126151391000	Chakok River 7.5 miles above mouth, near Anchor Point, Alaska
45	595506151403300	Stariski Creek 2 miles below unnamed Trail, near Ninilchik, Alaska
46	15240300	Stariski Creek near Anchor Point, Alaska
47	594507151290000	Beaver Creek 2 miles above mouth, near Bald Mountain, near Homer, Alaska
48	600107151112800	North Fork Deep Creek 4 miles above mouth, near Ninilchik, Alaska
49	601708154203500	Little Kijik River above Kijik Lake, near Port Alsworth, Alaska
50	601828154171700	Little Kijik River below Kijik Lake, near Port Alsworth, Alaska
51	601801154143600	Kijik River 1.5 miles above mouth, near Port Alsworth, Alaska
52	601833154154100	Kijik River above Little Kijik River, near Port Alsworth, Alaska
M8	15210600	Gilahina River at McCarthy Road, near Chitina, Alaska

Table 16. Stations used in non-metric multi-dimensional scaling analysis, Alaska.—Continued

[USGS, U.S. Geological Survey]

Site No.	USGS station No.	Site name
M7	15210300	Long Lake Creek at McCarthy Road, near McCarthy, Alaska
M3	15210260	Long Lake Tributary 1,150 feet above mouth, near McCarthy, Alaska
56	631018149323700	Costello Creek near Colorado, Alaska
57	631629149352000	Colorado Creek near Colorado, Alaska
58	15294700	Johnson River above Lateral Glacier, near Tuxedni Bay, Alaska
59	15292304	Costello Creek below Camp Creek, near Colorado, Alaska
60	15292302	Camp Creek at mouth, near Colorado, Alaska
61	15294100	Deshka River near Willow, Alaska
62	15273020	Rabbit Creek at Hillside Drive, near Anchorage, Alaska
63	15273100	Little Rabbit Creek near Anchorage, Alaska
64	15241600	Ninilchik River at Ninilchik, Alaska
65	15275100	Chester Creek at Arctic Boulevard, at Anchorage, Alaska
66	15274000	South Fork Campbell Creek near Anchorage, Alaska
M4	15210260	Long Lake Tributary 1,150 feet above mouth, near McCarthy, Alaska
M11	15210700	Chokosna River at McCarthy Road, near Chitina, Alaska
M9	15210600	Gilahina River at McCarthy Road, near Chitina, Alaska
70	15276570	Ship Creek below Power Plant, at Elmendorf AFB, Alaska
71	15276200	Ship Creek at Glenn Hwy, near Anchorage, Alaska
72	15273030	Rabbit Creek at East 140 Avenue, near Anchorage, Alaska
73	15273040	Rabbit Creek at Porcupine Trail Road, near Anchorage, Alaska
74	15274830	South Branch of South Fork Chester Creek at Boniface Parkway, near Anchorage, Alaska
75	15275100	Chester Creek at Arctic Boulevard, at Anchorage, Alaska
76	15274395	Campbell Creek at New Seward Highway, near Anchorage, Alaska
77	15274557	Campbell Creek at Creek Street, near Anchorage, Alaska
78	15273090	Little Rabbit Creek at Nickleen Streeet, near Anchorage, Alaska
79	15273097	Little Rabbit Creek at Goldenview Drive, near Anchorage, Alaska
80	15274796	South Branch of South Fork Chester Creek at Tank Trail, near Anchorage, Alaska
M1	15210200	Lakina River near McCarthy, Alaska
S8	15547575	Slate Creek 2.1 miles above Eldorado Creek, near Kantishna, Alaska
G8	15547560	Glen Creek 150 feet above Moose Creek, near Kantishna, Alaska
R8	15547510	Rock Creek 4 miles above Bearpaw River, near Kantishna, Alaska
C18	15547520	Caribou Creek 3.4 miles above Crevice Creek, near Kantishna, Alaska
E8	15547570	Eureka Creek near Kantishna, Alaska
C58	15547530	Caribou Creek 2.6 miles above Bearpaw River, near Kantishna, Alaska
F8	15547590	Friday Creek near Kantishna, Alaska
E9	15547570	Eureka Creek near Kantishna, Alaska
C19	15547520	Caribou Creek 3.4 miles above Crevice Creek, near Kantishna, Alaska
C59	15547530	Caribou Creek 2.6 miles above Bearpaw River, near Kantishna, Alaska
S9	15547575	Slate Creek 2.1 miles above Eldorado Creek, near Kantishna, Alaska
G9	15547560	Glen Creek 150 feet above Moose Creek, near Kantishna, Alaska
R9	15547510	Rock Creek 4 miles above Bearpaw River, near Kantishna, Alaska
F9	15547590	Friday Creek near Kantishna, Alaska

Algae

Samples were analyzed for all periphytic algae: diatoms, Phylum Chrysophyta; green, Phylum Chlorphyta; red, Phylum Rhodophyta; blue-green, Phylum Cyanophyta; and yellow, Phylum Chrysophyta, Class Chrysophyceae. A total of 164 taxa were documented at the Kantishna Hills sites (appendix H). Algal taxa richness (number of taxa found in a sample) ranged from 41 taxa at the most downstream Caribou Creek site (site 6) to 19 taxa at Friday Creek (site 14) (table 17). The vast majority of taxa at all sites were diatoms, from 36 taxa at Caribou Creek (site 6) to 16 taxa at Friday Creek; the highest non-diatom taxa richness was 7 at Caribou Creek (site 6) and the lowest was 2 non-diatom taxa identified at Slate Creek.

The overall richness of the non-motile algal taxa (diatoms), which account for 74–92 percent of algal taxa richness (table 17), suggests relatively stable streambeds at the sample sites. Motility of algae is considered to be a surrogate for disturbance in or adjacent to the stream. The ability to avoid being covered by sediments is a characteristic of motile algae as they are able to compensate for disturbances such as mining, agriculture, or urbanization that are known to increase the transport of fines in aquatic systems (Stevenson and others, 1996).

The two Slate Creek samples had the greatest percentage of abundance (calculated as density) of diatom taxa (table 18).

The 2009 samples from Friday Creek and Caribou Creek (site 6) had the next greatest abundance percentage of diatoms at 20 and 25 percent, respectively. Eureka Creek (site 11) and Caribou Creek (site 2) had the lowest diatom percent abundances (0.8 and 1.5, respectively). The remaining samples ranged from 5 to 11 percent diatom abundance. Green algae were identified in six samples, but it was most notable for being identified in the 2008 and 2009 samples for Rock and Caribou Creek (site 6). Blue-green algae (also known as cyanobacteria), often are in substantial areal extents or in a high biomass (for example, an algae bloom) in large part because of their capability to fix nitrogen (N_2) (Stevenson and others, 1996). This ability increases the number of potential habitats blue-green algae can inhabit because nitrogen is not a limiting growth factor, as it may be in the environment. The algal composition of the two Eureka Creek samples was almost entirely blue-green algae (92 and 99 percent). Slate Creek (2008 sample) and Caribou Creek (site 6, 2009 sample) had the lowest abundances of blue-green algae (23 and 43 percent, respectively). Red algae were identified in four samples; it consists of over a one-fourth of the abundance percentage for Slate Creek (2008 sample) and Rock Creek (2009 sample), and comprised 6 and 12 percent of algal abundance at Glen Creek (2009 sample) and at Caribou Creek (site 6, 2009 sample), respectively. Yellow algae was identified in only one sample, Glen Creek (2008 sample) (7 percent abundance), though it is relatively common in Alaska.

Table 17. Algae taxa richness by algae type, Kantishna Hills, Alaska, 2008–09.

[**Site No.:** Locations of sites shown in figure 2. –, non present]

Site No.	USGS station No.	Site name	Collection date	Taxa richness	Diatom	Green algae	Blue-green algae	Red algae	Yellow algae
1	15547510	Rock Creek	08-23-2008	39	35	1	3	–	–
			08-12-2009	32	27	2	2	1	–
2	15547520	Caribou Creek	08-24-2008	29	25	1	3	–	–
			08-11-2009	32	29	–	3	–	–
6	15547530	Caribou Creek	08-23-2008	41	36	2	3	–	–
			08-11-2009	37	30	4	2	1	–
9	15547560	Glen Creek	08-22-2008	31	27	1	2	–	1
			08-12-2009	26	22	–	3	1	–
11	15547570	Eureka Creek	08-24-2008	38	34	–	4	–	–
			08-10-2009	32	29	–	3	–	–
12	15547575	Slate Creek	08-21-2008	28	26	–	1	1	–
			08-11-2009	34	32	–	2	–	–
14	15547590	Friday Creek	08-24-2008	19	16	–	3	–	–
			08-11-2009	28	24	–	4	–	–

Table 18.   Abundance percentage of algae by type, Kantishna Hills, Alaska, 2008–09.

[**Site No.:** Locations of sites shown in figure 2. Number of organisms per unit area expressed as a percentage. –, none present]

Site No.	USGS station No.	Site name	Collection date	Diatoms	Non-diatoms	Green algae	Blue-green algae	Red algae	Yellow algae
1	15547510	Rock Creek	08-23-2008	6	94	15	79	–	–
			08-12-2009	11	89	4	57	28	–
2	15547520	Caribou Creek	08-24-2008	10	90	5	85	–	–
			08-11-2009	1	99	–	99	–	–
6	15547530	Caribou Creek	08-23-2008	9	91	9	82	–	–
			08-11-2009	25	75	20	43	12	–
9	15547560	Glen Creek	08-22-2008	12	88	3	78	–	7
			08-12-2009	5	95	–	89	6	–
11	15547570	Eureka Creek	08-24-2008	8	92	–	92	–	–
			08-10-2009	1	99	–	99	–	–
12	15547575	Slate Creek	08-21-2008	50	50	–	23	27	–
			08-11-2009	30	70	–	70	–	–
14	15547590	Friday Creek	08-24-2008	6	94	–	94	–	–
			08-11-2009	20	80	–	80	–	–

Four taxa, *Homoeothrix janthina* (blue-green algae), *Phormidium granulatum* (blue-green algae), *Batrachospermum sp.* (red algae), and *Geitlerinema splendidum* (blue-green algae), were the most dominant taxa as determined by the percentage of total abundance expressed as density (table 19). *Phormidium spp.* were the most dominant taxa in seven samples, and four of these samples represented the highest percentage of the most dominant taxa in the study. *Homoeothrix janthina* was the most dominant taxa in five samples, followed by one sample each for *Bactrachospermum sp.* and *Geitlerinema splendidum*.

The diversity of the algal community also was analyzed to add depth to the qualitative determination of taxa richness. Shannon diversity (Ludwig and Reynolds, 1988) incorporates the number of taxa in a sample combined with a weighted measure of abundance data. Larger values represent a greater diversity of taxa. The greatest diversity (fig. 28) of algae taxa was in Slate Creek (2008 sample; 1.02) followed by Caribou Creek (site 6) (2009 sample; 0.97). The lowest diversity was in Eureka Creek and Caribou Creek (site 2) (2009 sample; 0.11 and 0.10, respectively). All other diversity samples ranged from 0.74 to 0.50. Caribou Creek (site 2) and Eureka Creek had the greatest differences in diversity (0.72 and 0.81, respectively) between years. The lowest difference in diversity values were 0.06 and 0.07 recorded for Glen and Rock Creeks. The remaining sites ranged from 0.26 to 0.29 difference in diversity values.

Algae data were analyzed further using the ordination technique (NMDS). Algal density values were standardized for the lowest identifiable taxonomic level for each sample in order to generate a Bray-Curtis dissimilarity input matrix. The first dimension in ordination space of the algal community (fig. 29) indicates a definitive separation of samples by the year sampled. This is likely related to the precipitation that occurred prior to sample collection for algae and macroinvertebrates in 2008 compared to the low-precipitation warm summer of 2009. In 2008, higher streamflow increased the probability of stream bed scouring, leaving a reduced periphytic algal community in richness and abundance. Conversely, in 2009, optimal weather conditions prior to sample collection may have contributed to a general reversal of the algal community richness and abundance observed in 2008. Although other factors obviously contribute to the presence and absence of taxa, weather appears to have played a substantial role in the shaping of algal communities in 2008–09. The second dimension, though not as evident as the first dimension, contains groups that provide some insight into the sites and samples. As the 2008 samples were collected under elevated flow conditions relative to 2009, the algal densities at three sites (Slate Creek, Friday Creek, and Caribou Creek [site 6]) grouped together (fig. 29).

Table 19. Two most dominant taxa by abundance percentage, Kantishna Hills, Alaska, 2008–09.

Site No.	USGS station No.	Site name	Collection date	Most dominant taxa	Percent dominance	Second most dominant taxa	Percent dominance
1	15547510	Rock Creek	08-23-2008	*Homoeothrix janthina*	59	*Phormidium* sp.	19
			08-12-2009	*Homoeothrix janthina*	55	Unknown Rhodophyte Florideophycidae	28
2	15547520	Caribou Creek	08-24-2008	*Phormidium* sp.	74	*Pleurocapsa minor*	6
			08-11-2009	*Phormidium granulatum*	96	*Homoeothrix janthina*	2
6	15547530	Caribou Creek	08-23-2008	*Phormidium granulatum*	45	*Symplocastrum pencillatum*	25
			08-11-2009	*Homoeothrix janthina*	32	Unknown Rhodophyte Florideophycidae	12
9	15547560	Glen Creek	08-22-2008	*Phormidium* sp.	61	*Homoeothrix janthina*	17
			08-12-2009	*Phormidium granulatum*	58	*Homoeothrix janthina*	20
11	15547570	Eureka Creek	08-24-2008	*Homoeothrix janthina*	54	*Phormidium* sp.	29
			08-10-2009	*Phormidium granulatum*	95	*Homoeothrix janthina*	2
12	15547575	Slate Creek	08-21-2008	*Batrachospermum* sp.	27	*Homoeothrix janthina*	23
			08-11-2009	*Homoeothrix janthina*	44	*Phormidium* sp.	26
14	15547590	Friday Creek	08-24-2008	*Geitlerinema splendidum*	50	*Phormidium granulatum*	38
			08-11-2009	*Phormidium granulatum*	31	Unknown Cyanophyte (colonial coccoid)	30

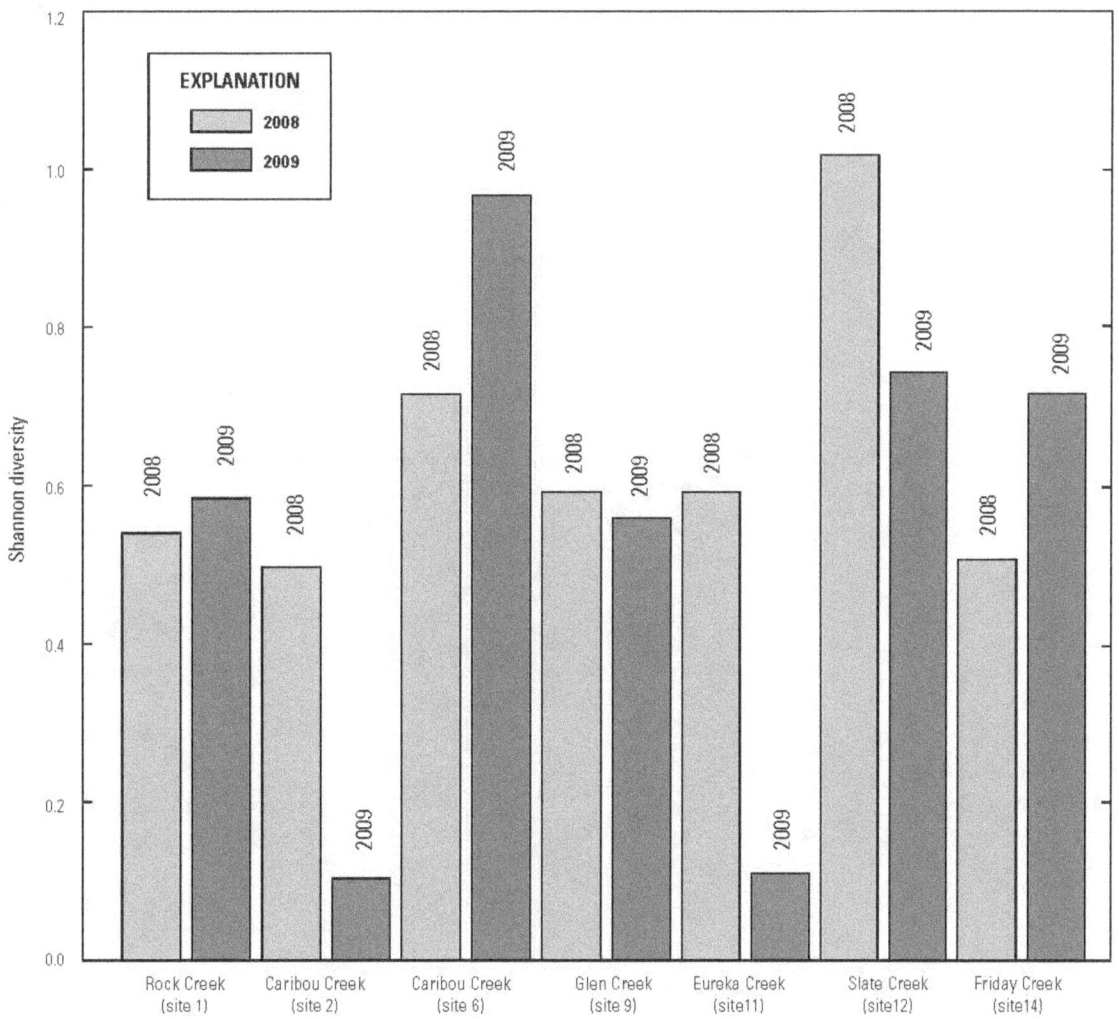

Figure 28. Shannon diversity for each algae sample from Kantishna Hills, Alaska, 2008–09.

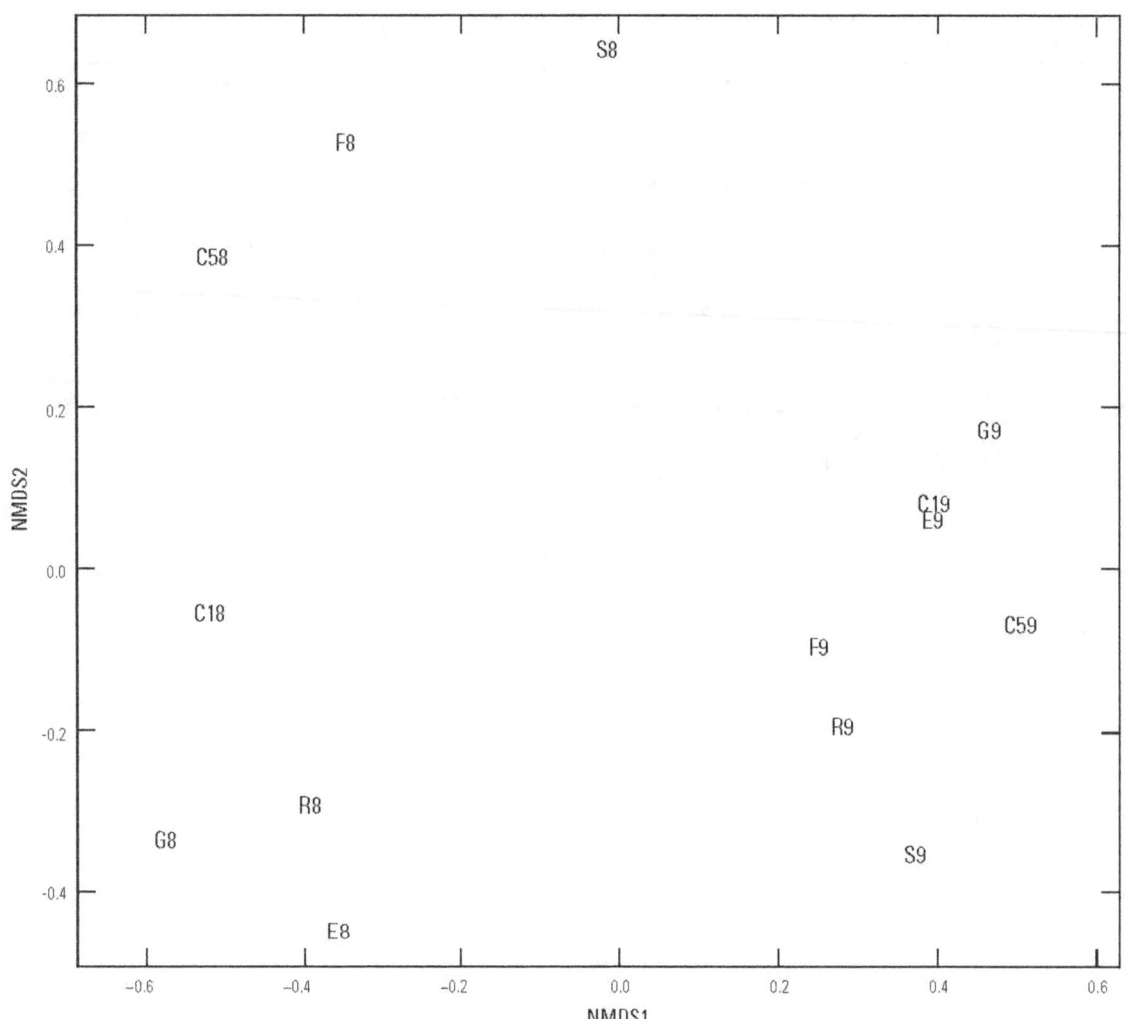

Figure 29. Non-metric multi-dimensional scaling (NMDS) ordination for algae samples collected from Kantishna Hills, Alaska, 2008–09. Site numbers are given in table 16.

Summary

The Kantishna Hills, in the northwest part of Denali National Park and Preserve, have undergone extensive mining that has resulted in water quality and aquatic habitat degradation to a number of streams. All mining in Kantishna Hills ceased in 1985 and since then the park has implemented a number of reclamation projects to improve the water quality and aquatic habitat of these streams. From 2008 to 2011, the U.S. Geological Survey (USGS) assessed the water quality conditions of streams in Kantishna Hills and the effects of reclamation work on water quality and aquatic habitat. Streams studied were Rock, Caribou, Glen, Slate, Eldorado, Eureka, Friday, and Moose Creeks. With the exception of Rock Creek, mining has occurred in these watersheds. Major findings of the study include:

- Since 1985, when mining ceased, turbidity and suspended-sediment concentrations have decreased. Data collected in the mid-1980s indicated high levels on streams where active mining occurred. During this study, turbidity and suspended-sediment concentrations were frequently less than 2 Formazin Nephelometric Units and 1 milligram per liter, respectively. Average daily turbidity collected in Rock Creek, an unmined stream, and in Caribou Creek, a mined stream, showed almost identical patterns in turbidity, an indication that Caribou Creek has improved both from reclamation efforts and through natural revegetation.

- Specific conductance and concentrations of dissolved solids were highest from streams that have been mined: Caribou Creek, Glen Creek, Eldorado Creek, Eureka Creek, and Friday Creek. With the exception of Caribou Creek, these streams flow into Moose Creek, which dilutes the specific conductance and ion concentrations of these streams. Alkalinity concentrations ranged from 49 to 194 milligrams per liter, indicating a high buffering capacity of the streams studied.

- Analyses of trace elements in samples of the water column indicated that concentrations of antimony at Slate Creek and Eureka Creek exceeded U.S. Environmental Protection Agency (USEPA) guidelines for drinking water. Arsenic concentrations at Slate Creek exceeded USEPA guidelines for drinking water in about one-half of the samples. Iron concentrations at Slate Creek exceed the threshold concentrations for chronic toxicity for aquatic life. Antimony and arsenic concentrations in streambed sediments were 5,800 and 3,900 micrograms per gram at Slate Creek, and 490 and 690 micrograms per gram at Eldorado Creek, respectively. Concentrations of arsenic, cadmium, chromium, copper, lead, nickel, and zinc in streambed sediments exceeded sediment quality guideline thresholds that could be toxic to aquatic life. Probable effect concentration quotients greater than 0.5, indicating potential toxicity to aquatic life, were detected at all streams except for Rock Creek and two sites on Moose Creek. Assessing these concentrations with the level of organic carbon indicate that only concentrations of arsenic and chromium may be toxic to aquatic life.

- In 2008–09, 104 macroinvertebrate taxa were identified from collected samples. Eighty-six percent of the macroinvertebrates were insects. Dipterans (flies and midges) accounted for 49 percent of all aquatic macroinvertebrates identified at the seven sites. Fourteen taxa, not found in other USGS sampled sites using National Water-Quality Assessment Program protocols, were identified in the samples collected in the Kantishna Hills. Worms and the stonefly (*Zapada oregonensis*) were identified in all samples.

- Based on the National Invertebrate Community Index, Rock Creek, a reference site, and the farthest downstream site on Caribou Creek, a mined stream that has undergone some reclamation, exhibited the best overall stream condition, whereas Slate Creek and Friday Creek, two small streams that were extensively mined, exhibited the worst stream condition.

- Compared to other sites in Alaska sampled by the USGS, 6 out of a possible 11 metrics for macroinvertebrate data for the sites sampled in the Kantishna Hills were determined to be statistically different. Five of these significantly different metrics suggest that the sites, with the exception of Slate Creek and Friday Creek, exhibit macroinvertebrate structure associated with minimal disturbance.

- Analysis of the macroinvertebrate data by non-metric multi-dimensional scaling (NMDS) showed a separation of sites by the different summers in 2008 (wet, cool) and in 2009 (warm, dry). A second NMDS of the Kantishna Hills macroinvertebrate data with other Alaska macroinvertebrate data collected by the USGS showed a distinct grouping between data from interior Alaska and data from southcentral Alaska, suggesting a distinction from continental type climate and transitional or maritime type climate.

- In 2008–09, 164 algae taxa were identified from collected samples. Most algal taxa were in the diatom group, followed by blue-green, green, red, and yellow algae. Blue-green taxa were the most dominant taxa. Precipitation likely played a role in algal community richness and abundance relative to bed scouring during high streamflow.

- A NMDS analysis of the algae taxa showed a distinct grouping between the 2008 and 2009 samples, most likely because of the differences in weather, similar to the grouping seen with macroinvertebrate data.

Acknowledgments

The authors wish to thank Guy Adema of the National Park Service for his support of this study. Field assistants from Denali National Park and Preserve who helped collect data included: Larissa Yocum, Nat Wilson, Denny Capps, Rob Burrows, Lilli Agel, Nadine Reitman, Jennifer Johnson, and Corrie Lane.

References Cited

Adler, Daniel, 2005, vioplot—Violin plot, R package version 0.2: Software Web site, accessed January 24, 2013, at http://neoscientists.org/~plex/.

Aiken, G.R., and Cotsaris, E., 1995, Soil and hydrology—Their effect on NOM: American Water Works Association, p. 26–45.

Alaska Department of Environmental Conservation, 2006, Alaska's Final 2006 Integrated Water Quality Monitoring and Assessment Report: Alaska Department of Environmental Conservation, accessed December 5, 2012, at http://www.dec.state.ak.us/water/wqsar/waterbody/docs/2006finalIntegratedReport.pdf.

Arbogast, B.F., ed., 1990, Quality assurance manual for the Branch of Geochemistry Survey: U.S. Geological Survey Open-File Report 90–668, 184 p.

Brabets, T.P., Nelson, G.L., Dorava, J.M., and Milner, A.M., 1999, Water-quality assessment of the Cook Inlet Basin, Alaska-Environmental Setting: U.S. Geological Survey Water-Resources Investigations Report 99–4025, 65 p.

Brabets, T.P., and Whitman, M.S., 2002, Water quality of Camp Creek, Costello Creek, and other selected streams on the south side of Denali National Park and Preserve, Alaska: U.S. Geological Survey Water-Resources Investigations Report 02–4260, 52 p.

Bray, J.R., and Curtis, J.T., 1957, An ordination of the upland forest communities of Southern Wisconsin: Ecology Monograph, v. 57, p. 325–349.

Bundtzen, T.K., 1981, Geology and mineral deposits of the Kantishna Hills, Mount McKinley quadrangle, Alaska: Fairbanks, Alaska, University of Alaska Fairbanks, Master's thesis, 476 p.

Canadian Council of Ministers of the Environment, 1999, Canadian sediment quality guidelines for the protection of aquatic life-Summary tables, in Canadian Environmental Quality Guidelines, 1999: Winnipeg, Canadian Council of Ministers of the Environment.

Crawford, J.K., and Luoma, S.N., 1993, Guidelines for studies of contaminants in biological tissues for the National Water-Quality Assessment Program: U.S. Geological Survey Open-File Report 92–494, 69 p.

Cuffney, T.F., and Brightbill, R.A., 2011, User's manual for the National Water-Quality Assessment Program Invertebrate Data Analysis System (IDAS) software, version 5: U.S. Geological Survey Techniques and Methods 7-C4, 126 p. (Also available at http://pubs.usgs.gov/tm/7c4/.)

Cuffney, T.F., Gurtz, M.E., and Meador, M.R., 1993, Methods for collecting benthic invertebrate samples as part of the National Water-Quality Assessment Program: U.S. Geological Survey Open-File Report 93–406, 66 p.

Deacon, J.R., and Stephens, V.C., 1998, Trace elements in streambed sediment and fish liver at selected sites in the Upper Colorado River Basin, Colorado, 1995–96: U.S. Geological Survey Water-Resources Investigations Report 98–4124, 19 p.

Densmore, R.V., Vander Meer, M.E., and Dunkle, N.G, 2000, Native plant revegetation manual for Denali National Park and Preserve: U.S. Geological Survey Information and Technology Report 2000–0006, 51 p.

Deschu, N.A., 1985a, Arsenic in sediments, water, and benthic fish in placer-mined and unmined streams, Denali National Park and Preserve, Alaska: University of Washington, Department of Civil Engineering, Master's thesis, 90 p.

Deschu, N.A., 1985b, Turbidity and settable solids in mined and unmined streams in the Kantishna Hills, Denali National Park and Preserve, Alaska: Research/Resources Management Report AR-7, 121 p.

Deschu, N.A., 1986, Water-related research and monitoring in the Kantishna Hills Mining District, Denali National Park and Preserve, Alaska: Anchorage, U.S. Department of the Interior, National Park Service, 100 p.

Deschu, N.A., and Kavanagh, R.C., 1986, Water quality and the effects of mining activities in the Kantishna Hills, Denali National Park and Preserve, Alaska: Research/Resources Management Report AR-5, 109 p.

Edwards, P.J., and Tranel, M.J., 1997, Physical and chemical characterization of streams and rivers within Denali National Park and Preserve, Alaska: Final Report submitted to National Park Service and U.S. Department of Agriculture, U.S. Forest Service, 76 p.

Eklund, A.C., 2010, Beeswarm—The bee swarm plot, an alternative to stripchart: R package version 0.0.7, accessed December 5, 2012, at http://CRAN.R-project.org/package=beeswarm.

Eppinger, R.G., Briggs, P.H., Crock, J.G., Meier, A.L., Sutley, S.J., and Theodorakos, P.M., 2000, Environmental-geochemical study of the Slate Creek antimony deposit, Kantishna Hills, Denali National Park and Preserve, Alaska, in Wilson, F.H., and Galloway, J.P., eds., Studies by the U.S. Geological Survey in Alaska, 2000: U.S. Geological Survey Professional Paper 1662, p. 123–141. (Also available at http://pubs.usgs.gov/pp/pp1662/.)

Fishman, M.J., 1993, Methods of analysis by the U.S. Geological Survey National Water-Quality Laboratory—Determination of inorganic and organic constituents in water and fluvial sediments: U.S. Geological Survey Open-File Report 93–125, 217 p.

Fishman, M.J., and Friedman, L.C., eds., 1989, Method for determination of inorganic substances in water and fluvial sediments: U.S. Geological Survey Techniques of Water-Resources Investigations, Book 5, Chap. A1, 545 p.

Gibbs, R.J., 1973, Mechanisms of trace metal transport in rivers: Science, v. 180, p. 71–73.

Hem, J.D., 1985, Study and interpretation of the chemical characteristics of natural water: U.S. Geological Survey Water-Supply Paper 2254, 264 p.

Hoffman, G.I., 1996, Methods of analysis by the U.S. Geological Survey National Water-Quality Laboratory-Preparation procedure for aquatic biological material determined for trace metals: U.S. Geological Survey Open-File Report 96–362, 42 p.

Horowitz, A.J., 1991, A primer on sediment-trace element chemistry (2d ed.): Chelsea, Michigan, Lewis Publishers, 136 p.

Hulten, E., 1968, Flora of Alaska and neighboring territories: Stanford, Calif., Stanford University Press, 1032 p.

Hynes, H.B.N., 1970 The ecology of running waters: Caldwell, New Jersey, reprint of first edition by Blackburn Press, 555 p.

Irons, J.G., Ray, S.R., Miller, L.K., and Oswood, M.W., 1989, Spatial and seasonal patterns of streambed water temperature in an Alaskan subarctic stream: Proceedings of the Symposium on Headwaters Hydrology, June 1989, Missoula, Montana, American Water Resources Association, Bethesda, Maryland, p. 381–390.

Karle, K.F., and Densmore, R., 1994, Stream and floodplain restoration in a riparian ecosystem disturbed by placer mining: Ecological Engineering, v. 3, p. 121–133.

Ludwig, J.A., and Reynolds, J.F., 1988, Statistical ecology—A primer on methods of computing: New York, John Wiley and Sons, 337 p.

Madison, R.J., 1981, Effects of placer mining on hydrologic systems in Alaska—State of knowledge: U.S. Geological Survey Open-File report 81–217, 25 p.

MacDonald, D.D., Ingersoll, C.G., and Berger, T.A., 2000, Development and evaluation of consensus-based sediment quality guidelines for freshwater ecosystems: Archives of Environmental Contamination and Toxicology, v. 39, p. 20–31.

Mangi Environmental Group, 2005, Water Resources Information and Issues Overview Report, Denali National Park and Preserve, Alaska: NPS/NRWRD/NRTR Technical Report 2005/341, 181 p.

Martin, J.M., and Whitfield, K., 1983, The significance of the river input of chemical elements to the ocean, in Wong, C.S., Boyle, E., Bruland, K.W., Burton, J.D., and Goldberg, E.D., eds., Trace metals in sea water: New York, Plenum Press, p. 127–138.

Meybeck, M., 1988, How to establish and use world budgets of riverine materials: in Lerman, A., and Meybeck, M., eds., Physical and chemical weathering in geochemical cycles: Dordrecht, The Netherlands, Kluwer Academic Publishers, p. 247–272.

Meyer, S.C., and Kavanagh, R.C., 1983, Fish resources and the effects of mining activities in the Kantishna Hills, Denali National Park, Alaska, 1982: U.S. Department of the Interior, National Park Service, Alaska Regional Office, 150 p.

National Park Service, 1990, Final environmental impact statement—Cumulative impacts of mining: National Park Service, Denali National Park and Preserve, v. 1–2, 14 p.

National Park Service, 1997, Denali National Park and Preserve North Access Feasibility Study: National Park Service Web site, accessed February 5, 2013, at http://www.nps.gov/dena/parkmgmt/northaccess.htm.

National Park Service, 2001, Environmental Assessment—Reclamation of mined lands program: National Park Service, Denali National Park and Preserve, 78 p.

Oksanen, Jari, Blanchet, F.G., Kindt, Roeland, Legendre, Pierre, O'Hara, R.B., Simpson, G.L., Solymos, Peter, Stevens, M.H.H., and Wagner, Helene, 2010, Vegan—Community ecology package: R package version 1.17-5, access December 5, 2012, http://CRAN.R-project.org/package=vegan.

Patton, C.J., and Truitt, E.P., 1992, Methods of analysis by the U.S. Geological Survey National Water-Quality Laboratory—Determination of total phosphorus by a kjeldahl digestion method and an automated colorimetric finish that includes dialysis: U.S. Geological Survey Open-File Report 92–146, 39 p.

Porter, S.D., Cuffney, T.F., Gurtz, M.E., and Meador, M.R., 1993, Methods for collecting algal samples as part of the National Water-Quality Assessment Program: U.S. Geological Survey Open-File Report 93–409, 39 p.

R Development Core Team, 2010, R—A language and environment for statistical computing: Vienna, Austria, R Foundation for Statistical Computing, version 2.11.1, accessed December 5, 2012, at http://www.R-project.org.

Scrimgeour, G.J., Prowse, T.D., Culp, J.M., and Chambers, P.A., 1994, Ecological effects of river ice break-up—A review and perspective: Freshwater Biology, v. 32, p. 261–275.

Shelton, L.R., 1994, Field guidelines for collecting and processing stream-water samples for the National Water-Quality Assessment Program: U.S. Geological Survey Open-File Report 94–455, 42 p.

Shelton, L.R., and Capel, P.D., 1994, Guidelines for collecting and processing samples of streambed sediment for analysis of trace elements and organic contaminants for the National Water-Quality Assessment program: U.S. Geological Survey Open-File Report 94–458, 20 p.

Shulski, M., and Wendler, G., 2007, The climate of Alaska: University of Alaska Press, 214 p.

Smith, K.S., and Huyck, L.O., 1999, An overview of the abundance, relative mobility, bioavailability, and human toxicity of metals: Reviews in Economic Geology, v. 6A, p. 29–70.

Stevenson, R.J., Bothwell, M.L., and Lowe, R.L., eds., 1996, Algal ecology: San Diego, Calif., Academic Press, Inc., 753 p.

Thornsberry, V.V., McKee, C.J., and Salisbury, W.G., 1984, 1983 mineral resource studies in the Kantishna Hills and Dunkle Mine areas, Denali National Park and Preserve: U.S. Bureau of Mines, v. 1, 234 p.

Thorp, J.H., and Covich, A.P., eds., 1991, Ecology and classification of North American freshwater invertebrates: San Diego, Calif., Academic Press, Inc., 911 p.

Turnipseed, D.P., and Sauer, V.B., 2010, Discharge measurements at gaging stations: U.S. Geological Survey Techniques and Methods book 3, chap. A8, 87 p. (Also available at http://pubs.usgs.gov/tm/tm3-a8/).

U.S. Department of the Interior, 1998, Guidelines for interpretation of the biological effects of selected constituents in biota, water, and sediment: National Irrigation Water Quality Program Information Report No. 3, 198 p.

U.S. Geological Survey, variously dated, National field manual for the collection of water-quality data: U.S. Geological Survey Techniques of Water-Resources Investigations, book 9, chaps. A1–A9 (Also available at http://pubs.water.usgs.gov/twri9A).

Van Derveer, W.D., and Canton, S., 1997, Selenium sediment toxicity thresholds and derivation of water quality criteria for freshwater biota of western streams: Environmental Toxicology and Chemistry, v. 16, p. 1260–1268.

Venables, W.N., and Ripley, B.D., 2002, Modern applied statistics with S (4th ed.): New York, Springer, 495 p.

Viereck, L.A., and Little, E.L., Jr., 1972, Alaska trees and shrubs—Agriculture handbook 410: Washington, D.C., U.S. Forest Service, 265 p.

Wagner, R.J., Boulger, R.W., Oblinger, C.J., and Smith, B.A., 2006, Guidelines and standard procedures for continuous water-quality monitors—Station operation, record computation, and data reporting: U.S. Geological Survey Techniques and Methods 1-03, 51 p.

West, R.L., 1982, Kantishna Hills heavy metals investigation: Denali National Park Interagency Agreement 14-16-0007-82-5524, 38 p.

West, R.L., and Deschu, N.A., 1984, Kantishna Hills heavy metals investigations, Denali National Park, 1983: Fairbanks, Alaska, U.S. Fish and Wildlife Service/Park Service Report under Interagency Agreement No. 14-16-007-82-5524, 42 p.

Western Regional Climate Center, 2012, RAWS USA climate archive, Wonder Lake, Alaska: Western Regional Climate Center database, accessed November 27, 2012, at http://www.raws.dri.edu/cgi-bin/rawMAIN.pl?akAWON.

Appendixes

Appendixes are presented as Microsoft© Excel spreadsheets. They can be accessed and downloaded at http://pubs.usgs.gov/sir/2013/5048/.

Appendix A. Field Water-Quality Data from Kantishna Hills, Alaska, 1982

Appendix B. Trace Element Water-Quality Data from Kantishna Hills, Alaska, 1982

Appendix C. Field Water-Quality Data from Kantishna Hills, Alaska, 1983

Appendix D. Trace Element Water-Quality Data from Kantishna Hills, Alaska, 1983

Appendix E. Turbidity and Streamflow Data from Kantishna Hills, Alaska, 1981–85

Appendix F. Water-Quality Data from Kantishna Hills, Alaska, 1994–96

Appendix G. Benthic Macroinvertebrate Community Detected at Stream Sites in Kantishna Hills, Alaska, 2008–09

Appendix H. Algae Taxa Detected at Stream Sites in Kantishna Hills, Alaska, 2008–09

www.ingramcontent.com/pod-product-compliance
Lightning Source LLC
Chambersburg PA
CBHW081600170526

45166CB00009B/2765